The New Era of Space Stations

Copyright Page

This book is copyrighted for 2022

Title: The New Era of Space Stations

The Living in Space Series Book 10

By Martin K. Ettington

All Rights Reserved USA 2022

ISBN: 9798815402447

Printed in the United States of America

… The New Era of Space Stations

The New Era of Space Stations

Other books by Martin K. Ettington

Spiritual and Metaphysics Books:
Prophecy: A History and How to Guide
God Like Powers and Abilities
Enlightenment for Newbies
Removing Illusions to Find True Happiness
Using the Scientific Method to Study the Paranormal
A Compendium of Metaphysics and How to Guides (Six books together in one volume)
Love from the Heart
The Enlightenment Experience
Learn Your Soul's Purpose
Pursuing Enlightenment
A Modern Man's Search for Truth
Use Intuition and Prophecy to Improve Your Life
The Handbook of Spiritual and Energy Healing
Pure Spirituality and God
Memories Before Birth and Reincarnation
Paranormal Abilities and the Yoga Sutras of Patanjali
Mystical and Magical Societies and Practitioners

Longevity & Immortality:
Physical Immortality: A History and How to Guide
The Commentaries of Living Immortals
Records of Extremely Long Lived Persons
Enlightenment and Immortality
Longevity Improvements from Science
The 10 Principles of Personal Longevity
Telomeres & Longevity
The Diets and Lifestyles of the World's Oldest Peoples
The Longevity Six Books Bundle
Long Lived Plants and Animals

Science Fiction:
Out of This Universe
The Immortals of the Interstellar Colony
The Mystic Soldier
The Immortality Sci Fi Bundle
Visiting Many Universes
The History of Science Fiction and Fantasy

The God Like Powers Series:
Human Invisibility
Invulnerability and Shielding
Teleportation
Psychokinesis
Our Energy Body, Auras, and Thoughtforms
The God Like Powers Series—Volume 1 Compilation

The Yoga Discovery Series:
Yoga-An Ancient Art Form
Hatha Yoga-Helping you Live Better
Raja Yoga-Through the Ages
The Yoga Discovery Package

Business & Coaching Books:
Creating, Paublishing, & Marketing Practitioner Ebooks
Building a Successful Longevity Coaching Business
Why Become a Coach?
The Professional Coaching Success Trilogy
2020-Make Money Writing and Selling Books
The 2020 Handbook of High Paying Work Without a College Degree
The important of Creativity and How to Improve Yours
Quantum Mechanics, Technology, Consciousness, and the Multiverse

Self-Improvement
Stress Relief and Methods to do so
The Importance of Creativity and How to Improve Yours
Building Self-Confidence
See the World Clearly
A Trilogy of Self Help Books

The New Era of Space Stations

A New Paradigm of Truth and Happiness

Science, Technology, and Misc.
Future Predictions By and Engineer & Seer
The Unusual Science & Technology Bundle

Removing Limits On Our Consciousness-And Thinking Outside the Box
Universal Holistic Philosophy
Ball Lightning
Stranger Than Science Stories and Facts
Planet Earth is Conscious

Survival
Survival of Humanity Throughout the Ages
33 Incredible True Survival Stories
The Importance of Fire in History and Mythology
How to Survive Anything: From the Wilderness to Man Made Disasters
Building and Stocking a Nuclear Shelter for less than $10,000
The Human Survival Five Books Bundle
Stranger Than Science Facts and Stories
Stranger Than Science Facts and Stories Volume Two

Legendary Beings
Are Cryptozoological Animals Real or Imaginary?
Fire in History and Mythology
All About Dragons
Sea Serpents and Ocean Monsters
The Legendary Animals Five Books Bundle
The Mythical People of Ireland
Bigfoot Mysteries and Some Answers
About the Little People: Fairies, Elves, Dwarfs and Leprechauns

Ancient History
The Real Atlantis-In the Eye of the Sahara
Ancient & Prehistoric Civilizations
Ancient & Prehistoric Civilizations-Book Two
The History of Antediluvian Giants
The Antediluvian History of Earth
Ancient Underground Cities and Tunnels
Strange Objects Which Should Not Exist
More Out of Place Artifacts
Strange and Ancient Places in the USA
A Theory of Ancient Prehistory And Giant Aliens
The Destruction of Civilization About 10,500 B.C.
A Timeline of Intelligent Life on Earth
A 300 Million Year Old Civilization Existed on Earth
The Encyclopedia of Out of Place Artifacts

Aliens and Space
Aliens and Secret Technology
Aliens Are Already Among Us
Designing and Building Space Colonies
Humanity and the Universe
All About Moon Bases
All About Mars Journeys and Settlement
The Space and Aliens Six Books Bundle
A Theory of Ancient Prehistory and Giant Aliens
The Space Colonies and Space Structures Coloring Book
All About Asteroids
Spaceships, Past, Present, and Future
Astronauts, Cosmonauts, and Other Important Space Flyers
All About Mars Journeys and Settlement
Mining the Asteroid Belt

The New Era of Space Stations

<u>Time Travel and Dimensions</u>
Real Time Travel Stories From a Psychic Engineer
The Real Nature of Time: An Analysis of Physics, Prophecy, and Time Travel Experiences
Stories of Parallel Dimensions
We Live in a Malleable Reality-and We Can Change It
The Time, Dimensions, and Quantum Mechanical Bundle
Alternate Dimensions & the Otherworld

<u>Political and Social</u>

The Empire of the United States: Forged By God's Spirit Through Man

The New Era of Space Stations

The Longevity Training Series

(A transcription of the online Multimedia Longevity Coaching Training Program)

The Personal Longevity Training Series-Book1-Long Lived Persons
The Personal Longevity Training Series-Book2-Your Soul's Purpose
The Personal Longevity Training Series-Book3-Enable Your Life Urge
The Personal Longevity Training Series-Book4-Your Spiritual Connection
The Personal Longevity Training Series-Book5-Having Love in Your Heart
The Personal Longevity Training Series-Book6-Energy Body Health
The Personal Longevity Training Series-Book7-The Science of Longevity
The Personal Longevity Training Series-Book8-Physical Body Health
The Personal Longevity Training Series-Book9-Avoiding Accidents
The Personal Longevity Training Series-Book10-Implementing These Principles

The Personal Longevity Training Series-Books One Thru Ten

These books are all available in digital and printed formats from my
website and on Amazon, Barnes & Noble, Apple ITunes, and many other sites

My Books Website is: http://mkettingtonbooks.com

The New Era of Space Stations

Signup for our Mailing List to get the following:

1) A discount coupon for 25% discount on all books on our site
2) Occasional Notices of new books available
3) Occasional Email on other offerings of ours (Monthly)

If you have any questions about this book or other subjects please contact the Author at:

mke@mkettingtonbooks.com

The New Era of Space Stations

The New Era of Space Stations

Table of Contents

1.0 Introduction 1
2.0 Past Space Stations 3
  2.1 The Soviet Salyut One 3
  2.2 The Soviet MIR 7
  2.3 United States Skylab..11
3.0 The International Space Station 15
4.0 New Space Transportation Options.......................19
  4.1 USA-Space X Dragon Manned 19
  4.2 USA-Boeing Starliner...23
  4.3 USA-NASA Orion..27
  4.4 USA-Sierra Nevada-Dream Chaser 31
5.0 Space Stations Being Planned Now......................33
  5.1 Orbital Reef Space Station.................................39
  5.2 Northrop Grumman's Space Station 45
  5.3 Nanoracks Space Station 49
  5.4 Pioneer Space Station 53
6.0 New Technologies for Space Stations........................57
  6.1 Inflatable Structures 57
  6.2 Growing Food in Space 61
  6.3 Working in Space...67
7.0 New Space Station Usage Plans...........................69
8.0 The Deep Space Gateway.....................................81
9.0 Latest Moon Base Plans...87
10.0 Russian Space Station Plans 91

The New Era of Space Stations

- 11.0 Chinese Space Stations 93
  - 11.1 China's Current Space Station 93
  - 11.2 China's Plans for a Giant Space Station 97
- 12.0 Future Advanced Space Colonies 103
- 13.0 Summary 113
- 14.0 Bibliography 115

The New Era of Space Stations

1.0 Introduction

We are at an exciting threshold for the development and deployment of private space stations and more advanced designs from various countries. This trend shows how quickly space technologies are developing.

The advancements in cheaper rockets to carry more payloads into Earth's orbit in a less costly manner is another trend which is supporting the development of these new space stations and space transportation options.

The last few years a lot of companies and governments are making plans to develop and deploy a variety of space stations and moon bases. This seems like to good time to update everyone with what are these latest plans, space technologies, transportation options and more.

I've tried to include all I could find on new space technologies and plans relevant to working and living in space.

I grew up in the nineteen sixties when the Space Race with the Russians was just starting. Walking to elementary school with my friends we used to pretend we were the Mercury Astronauts.

Later in the nineteen eighties I worked at the Johnson Space Center and with Contractors for a couple of years. I even applied to the Astronaut program once. So I love Space Travel.

This led me to write my first book on Space in 2017 titled "Designing and Building Space Colonies" along with another ten space related books later on.

# The New Era of Space Stations

The New Era of Space Stations

2.0 Past Space Stations

The Soviets launched the first space station and had a number of others. The first US Space Station was Skylab. This chapter covers those most notable stations.

2.1 The Soviet Salyut One

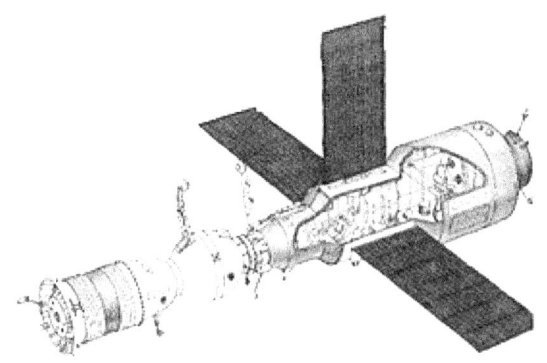

Salyut 1 (DOS-1) was the world's first space station launched into low Earth orbit by the Soviet Union on April 19, 1971. The Salyut program followed this with five more successful launches of seven more stations. The final module of the program, Zvezda (DOS-8), became the core of the Russian segment of the International Space Station and remains in orbit.

Salyut 1 was modified from one of the Almaz airframes, and was made out of five components: a transfer compartment, a main compartment, two auxiliary compartments, and the Orion 1 Space Observatory.

Salyut 1 was visited by Soyuz 10 and Soyuz 11. The hard-docking of Soyuz 10 failed and the crew had to abort this mission. The Soyuz 11 crew achieved successful hard docking and performed experiments in Salyut 1 for 23

The New Era of Space Stations

days. However, they were killed by asphyxia caused by failure of a valve just prior to Earth reentry, and are the only people to have died above the Kármán line. Salyut 1's mission was later terminated, and it burned up on reentry into Earth's atmosphere on October 11, 1971.

Structure

At launch, the announced purpose of Salyut was to test the elements of the systems of a space station and to conduct scientific research and experiments. The craft was described as being 20 m (66 ft) in length, 4 m (13 ft) in maximum diameter, and 99 m3 (3,500 cu ft) in interior space with an on-orbit dry mass of 18,425 kg (40,620 lb). Of its several compartments, three were pressurized (100 m³ total), and two could be entered by the crew.

Transfer compartment

The transfer compartment was equipped with the only docking port of Salyut 1, which allowed one Soyuz 7K-OKS spacecraft to dock. It was the first use of the Soviet SSVP docking system that allowed internal crew transfer, a system that is in use today. The docking cone had a 2 m (6.6 ft) front diameter and a 3 m (9.8 ft) aft diameter.

Main compartment

The second and main compartment was about 4 m (13 ft) in diameter. Televised views showed enough space for eight large chairs (seven at work consoles), several control panels, and 20 portholes (some obstructed by instruments). In Salyut 1 the interior design used various colors (light and dark gray, apple green, light yellow) for supporting the cosmonauts' orientation in weightlessness.

The New Era of Space Stations

Auxiliary compartments

The third pressurized compartment contained the control and communications equipment, the power supply, the life support system, and other auxiliary equipment. The fourth and final unpressurized compartment was about 2 m in diameter and contained the engine installations and associated control equipment. Salyut had buffer chemical batteries, reserve supplies of oxygen and water, and regeneration systems. Externally mounted were two double sets of solar cell panels that extended like wings from the smaller compartments at each end, the heat regulation system's radiators, and orientation and control devices.

Salyut 1 was modified from one of the Almaz airframes. The unpressurized service module was the modified service module of a Soyuz craft.

Orion 1 Space Observatory

The astrophysical Orion 1 Space Observatory designed by Grigor Gurzadyan of Byurakan Observatory in Armenia, was installed in Salyut 1. Ultraviolet spectrograms of stars were obtained with the help of a mirror telescope of the Mersenne system and a spectrograph of the Wadsworth system using film sensitive to the far ultraviolet. The dispersion of the spectrograph was 32 Å/mm (3.2 nm/mm), while the resolution of the spectrograms derived was about 5 Å at 2600 Å (0.5 nm at 260 nm). Slitless spectrograms were obtained of the stars Vega and Beta Centauri between 2000 and 3800 Å (200 and 380 nm). The telescope was operated by crew member Viktor Patsayev, who became the first man to operate a telescope outside of the Earth's atmosphere.

The New Era of Space Stations

The New Era of Space Stations

2.2 The Soviet MIR

A friend of mine David Wolf actually spent several months on the MIR space station.

Mir was a space station that operated in low Earth orbit from 1986 to 2001, operated by the Soviet Union and later by Russia. Mir was the first modular space station and was assembled in orbit from 1986 to 1996. It had a greater mass than any previous spacecraft. At the time it was the largest artificial satellite in orbit, succeeded by the International Space Station (ISS) after Mir's orbit decayed. The station served as a microgravity research laboratory in

The New Era of Space Stations

which crews conducted experiments in biology, human biology, physics, astronomy, meteorology, and spacecraft systems with a goal of developing technologies required for permanent occupation of space.

Mir was the first continuously inhabited long-term research station in orbit and held the record for the longest continuous human presence in space at 3,644 days, until it was surpassed by the ISS on 23 October 2010. It holds the record for the longest single human spaceflight, with Valeri Polyakov spending 437 days and 18 hours on the station between 1994 and 1995. Mir was occupied for a total of twelve and a half years out of its fifteen-year lifespan, having the capacity to support a resident crew of three, or larger crews for short visits.

Following the success of the Salyut program, Mir represented the next stage in the Soviet Union's space station program. The first module of the station, known as the core module or base block, was launched in 1986 and followed by six further modules. Proton rockets were used to launch all of its components except for the docking module, which was installed by US Space Shuttle mission STS-74 in 1995. When complete, the station consisted of seven pressurized modules and several unpressurised components. Power was provided by several photovoltaic arrays attached directly to the modules. The station was maintained at an orbit between 296 km (184 mi) and 421 km (262 mi) altitude and travelled at an average speed of 27,700 km/h (17,200 mph), completing 15.7 orbits per day.

The station was launched as part of the Soviet Union's crewed spaceflight program effort to maintain a long-term research outpost in space, and following the collapse of the USSR, was operated by the new Russian Federal Space Agency (RKA). As a result, most of the station's occupants were Soviet; through international collaborations such as the Intercosmos, Euromir and Shuttle–Mir programs, the station was made accessible to space travelers from several Asian, European and North American nations. Mir was deorbited in March 2001 after funding was cut off. The cost of the Mir program was estimated by former RKA General Director Yuri Koptev in 2001 as $4.2 billion over its lifetime (including development, assembly and orbital operation).

The New Era of Space Stations

The New Era of Space Stations

2.3 United States Skylab

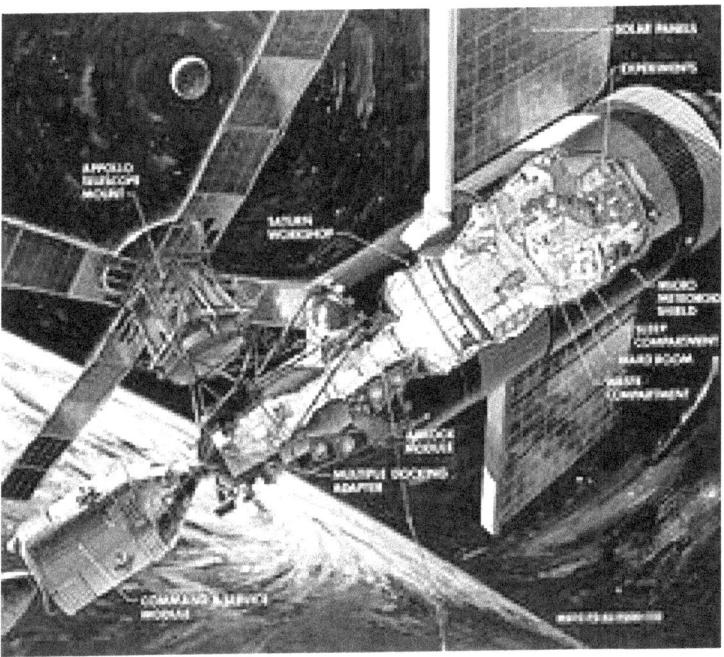

Skylab was the first United States space station, launched by NASA, occupied for about 24 weeks between May 1973 and February 1974. It was operated by three separate three-astronaut crews: Skylab 2, Skylab 3, and Skylab 4. Major operations included an orbital workshop, a solar observatory, Earth observation, and hundreds of experiments.

Unable to be re-boosted by the Space Shuttle, which was not ready until 1981, Skylab's orbit eventually decayed, and it disintegrated in the atmosphere on July 11, 1979, scattering debris across the Indian Ocean and Western Australia.

The New Era of Space Stations

Habitability

A dry workshop simplified plans for the interior of the station. Industrial design firm Raymond Loewy/William Snaith recommended emphasizing habitability and comfort for the astronauts by providing a wardroom for meals and relaxation and a window to view Earth and space, although astronauts were dubious about the designers' focus on details such as color schemes. Habitability had not previously been an area of concern when building spacecraft due to their small size and brief mission durations, but the Skylab missions would last for months. NASA sent a scientist on Jacques Piccard's Ben Franklin submarine in the Gulf Stream in July and August 1969 to learn how six people would live in an enclosed space for four weeks.

Astronauts were uninterested in watching movies on a proposed entertainment center or in playing games, but they did want books and individual music choices. Food was also important; early Apollo crews complained about its quality, and a NASA volunteer found it intolerable to live on the Apollo food for four days on Earth. Its taste and composition were unpleasant, in the form of cubes and squeeze tubes. Skylab food significantly improved on its predecessors by prioritizing palatability over scientific needs.

Each astronaut had a private sleeping area the size of a small walk-in closet, with a curtain, sleeping bag, and locker. Designers also added a shower and a toilet for comfort and to obtain precise urine and feces samples for examination on Earth. The waste samples were so important that they would have been priorities in any rescue mission.

The New Era of Space Stations

Skylab did not have recycling systems such as the conversion of urine to drinking water; it also did not dispose of waste by dumping it into space. The S-IVB's 73,280 liters (16,120 imp gal; 19,360 U.S. gal) liquid oxygen tank below the Orbital Work Shop was used to store trash and wastewater, passed through an airlock.

Operational history

On August 8, 1969, the McDonnell Douglas Corporation received a contract for the conversion of two existing S-IVB stages to the Orbital Workshop configuration. One of the S-IV test stages was shipped to McDonnell Douglas for the construction of a mock-up in January 1970. The Orbital Workshop was renamed "Skylab" in February 1970 as a result of a NASA contest. The actual stage that flew was the upper stage of the AS-212 rocket (the S-IVB stage, S-IVB 212). The mission computer used aboard Skylab was the IBM System/4Pi TC-1, a relative of the AP-101 Space Shuttle computers. The Saturn V with serial number SA-513, originally produced for the Apollo program – before the cancellation of Apollo 18, 19, and 20 – was repurposed and redesigned to launch Skylab. The Saturn V's third stage was removed and replaced with Skylab, but with the controlling Instrument Unit remaining in its standard position.

Skylab was launched on May 14, 1973, by the modified Saturn V. The launch is sometimes referred to as Skylab 1. Severe damage was sustained during launch and deployment, including the loss of the station's micrometeoroid shield/sun shade and one of its main solar panels. Debris from the lost micrometeoroid shield further complicated matters by becoming tangled in the remaining solar panel, preventing its full deployment and thus leaving the station with a huge power deficit.

The New Era of Space Stations

Immediately following Skylab's launch, Pad 39A at Kennedy Space Center was deactivated, and construction proceeded to modify it for the Space Shuttle program, originally targeting a maiden launch in March 1979. The crewed missions to Skylab would occur using a Saturn IB rocket from Launch Pad 39B.

Skylab 1 was the last uncrewed launch from LC-39A until February 19, 2017, when SpaceX CRS-10 was launched from there.

The New Era of Space Stations

3.0 The International Space Station

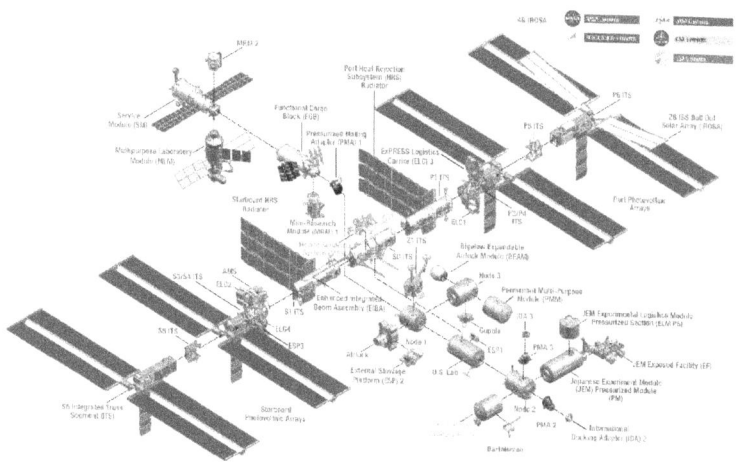

The International Space Station (ISS) is a modular space station (habitable artificial satellite) in low Earth orbit. It is a multinational collaborative project involving five participating space agencies: NASA (United States), Roscosmos (Russia), JAXA (Japan), ESA (Europe), and CSA (Canada). The ownership and use of the space station is established by intergovernmental treaties and agreements. The station serves as a microgravity and space environment research laboratory in which scientific research is conducted in astrobiology, astronomy, meteorology, physics, and other fields. The ISS is suited for testing the spacecraft systems and equipment required for possible future long-duration missions to the Moon and Mars.

The ISS program evolved from the Space Station Freedom, a 1984 American proposal to construct a permanently manned Earth-orbiting station, and the contemporaneous Soviet/Russian Mir-2 proposal from 1976 with similar aims. The ISS is the ninth space station

The New Era of Space Stations

to be inhabited by crews, following the Soviet and later Russian Salyut, Almaz, and Mir stations and the American Skylab. It is the largest artificial object in space and the largest satellite in low Earth orbit, regularly visible to the naked eye from Earth's surface. It maintains an orbit with an average altitude of 400 kilometers (250 mi) by means of reboost maneuvers using the engines of the Zvezda Service Module or visiting spacecraft. The ISS circles the Earth in roughly 93 minutes, completing 15.5 orbits per day.

The station is divided into two sections: the Russian Orbital Segment (ROS) is operated by Russia, while the United States Orbital Segment (USOS) is run by the United States as well as by the other states. The Russian segment includes six modules. The US segment includes ten modules, whose support services are distributed 76.6% for NASA, 12.8% for JAXA, 8.3% for ESA and 2.3% for CSA.

Roscosmos had endorsed the continued operation of ROS through 2024, having previously proposed using elements of the segment to construct a new Russian space station called OPSEK. However, continued cooperation has been rendered uncertain by the 2022 Russian invasion of Ukraine and subsequent international sanctions on Russia, who theoretically, may lower, redirect, or cut funding their side of the space station due to the sanctions set on them.

The first ISS component was launched in 1998, and the first long-term residents arrived on 2 November 2000 after being launched from the Baikonur Cosmodrome on 31 October 2000. The station has since been continuously occupied for 21 years and 179 days, the longest continuous human presence in low Earth orbit, having surpassed the previous record of 9 years and 357 days held by the Mir space station. The latest major pressurized

The New Era of Space Stations

module, Nauka, was fitted in 2021, a little over ten years after the previous major addition, Leonardo in 2011.

Development and assembly of the station continues, with an experimental inflatable space habitat added in 2016, and several major new Russian elements scheduled for launch starting in 2021. In January 2022, the station's operation authorization was extended to 2030, with funding secured through that year. There have been calls to privatize ISS operations after that point to pursue future Moon and Mars missions, with former NASA Administrator Jim Bridenstine stating: "given our current budget constraints, if we want to go to the moon and we want to go to Mars, we need to commercialize low Earth orbit and go on to the next step."

The ISS consists of pressurized habitation modules, structural trusses, photovoltaic solar arrays, thermal radiators, docking ports, experiment bays and robotic arms. Major ISS modules have been launched by Russian Proton and Soyuz rockets and US Space Shuttles. The station is serviced by a variety of visiting spacecraft: the Russian Soyuz and Progress, the SpaceX Dragon 2, and the Northrop Grumman Space Systems Cygnus, and formerly the European Automated Transfer Vehicle (ATV), the Japanese H-II Transfer Vehicle, and SpaceX Dragon 1.

The Dragon spacecraft allows the return of pressurized cargo to Earth, which is used, for example, to repatriate scientific experiments for further analysis. As of April 2022, 251 astronauts, cosmonauts, and space tourists from 20 different nations have visited the space station, many of them multiple times.

The New Era of Space Stations

The New Era of Space Stations

4.0 New Space Transportation Options

Spaceships used to launch cargo and crews into space are also advancing which provides more lower cost options to support manned space stations. Here are some of the most interesting vehicles:

4.1 USA-Space X Dragon Manned

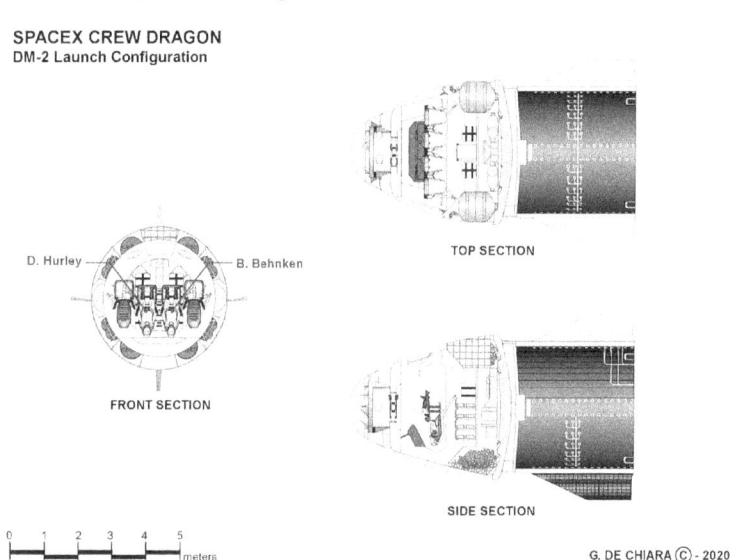

The SpaceX Dragon 2 is a class of reusable spacecraft developed and manufactured by American aerospace manufacturer SpaceX as the successor to Dragon, a reusable cargo spacecraft. It has two variants: Crew Dragon, a space capsule capable of ferrying up to seven astronauts, and Cargo Dragon, an updated replacement for the original Dragon spacecraft. The spacecraft launches atop a Falcon 9 Block 5 rocket and returns to Earth via an ocean splashdown. Unlike its predecessor, the spacecraft can dock itself to the ISS instead of being

The New Era of Space Stations

berthed. Crew Dragon is equipped with an integrated launch escape system (LES) capable of accelerating the vehicle away from the rocket in an emergency at 11.8 m/s2 (39 ft/s2), accomplished by using a set of four side-mounted thruster pods with two SuperDraco engines each. The spacecraft features redesigned solar arrays and a modified outer mold line compared to the original Dragon, and possesses new flight computers and avionics. As of March 2020, four Dragon 2 spacecraft have been manufactured (not counting structural test articles that were never airborne).

Crew Dragon serves as one of two spacecraft that is expected to transport crews to and from the International Space Station (ISS) under NASA's Commercial Crew Program, the other being the Boeing CST-100 Starliner. It is also expected to be used in flights by American space tourism company Space Adventures and to shuttle tourists to and from Axiom Space's planned space station. Crew Dragon's first non-piloted test flight occurred in March 2019, and its first crewed flight – with astronauts Robert Behnken and Douglas Hurley – occurred in May 2020. This test flight marked the first time a private company launched a crewed orbital spacecraft. Cargo Dragon is expected to supply cargo to the ISS under a Commercial Resupply Services-2 contract with NASA, along with Northrop Grumman Innovation Systems' Cygnus spacecraft and Sierra Nevada Corporation's Dream Chaser spacecraft. The first flight of the Cargo Dragon is planned to launch in October 2020.

There are two variants: Crew Dragon and Cargo Dragon. Crew Dragon was initially called "DragonRider" and it was intended from the beginning to support a crew of seven or a combination of crew and cargo. It is able to perform fully autonomous rendezvous and docking with manual override ability, using the NASA Docking System (NDS). For typical missions, Crew Dragon will remain docked to the ISS for a period of 180 days, but is designed to remain on the station for up to 210 days, matching the Russian Soyuz spacecraft. From the beginning of the development process, SpaceX planned to use an integrated pusher launch escape system for the Dragon spacecraft.

The SpaceX Dragon 2 Control Panel

The New Era of Space Stations

The New Era of Space Stations

4.2 USA-Boeing Starliner

Boeing Starliner (officially CST-100 Starliner) is a class of reusable crew capsules expected to transport crew to the International Space Station (ISS) and to private space stations such as the proposed Bigelow Aerospace Commercial Space Station. It is manufactured by Boeing for its participation in NASA's Commercial Crew Program.

The capsule has a diameter of 4.56 m (15.0 ft), which is slightly larger than the Apollo command module and smaller than the Orion capsule. The Boeing Starliner holds a crew of up to seven people and is being designed to be able to remain in-orbit for up to seven months with reusability of up to ten missions. It is designed to be compatible with four launch vehicles: Atlas V, Delta IV, Falcon 9, and Vulcan.

The New Era of Space Stations

In the first phase of its CCP, NASA awarded Boeing US$18 million in 2010 for preliminary development of the spacecraft. In the second phase Boeing was awarded a US$93 million contract in 2011 for further spacecraft development. On 3 August 2012, NASA announced the award of US$460 million to Boeing to continue work on the Starliner under the Commercial Crew Integrated Capability (CCiCap) Program. On 16 September 2014, NASA selected the Boeing Starliner, along with SpaceX Crew Dragon, for the Commercial Crew Transportation Capability (CCtCap) program, with an award of US$4.2 billion.

The Boeing Starliner Orbital Flight Test (uncrewed test flight) launched with the Atlas V N22, on 20 December 2019 from SLC-41 at Cape Canaveral, Florida. During the test, the Starliner experienced a timing anomaly that precluded a docking with the International Space Station. Two days after launch, on 22 December 2019 at 12:58 UTC, with the successful landing at White Sands Missile Range, New Mexico, the Boeing Starliner Calypso became the first-ever crew-capable space capsule to make a land-based touchdown in the United States.

The design draws upon Boeing's experience with NASA's Apollo, Space Shuttle and ISS programs as well as the Orbital Express project sponsored by the Department of Defense. Starliner has no Orion heritage, but it is sometimes confused with the earlier and similar Orion-derived Orion Lite proposal that Bigelow Aerospace was reportedly working on with technical assistance from Lockheed Martin. It will use the NASA Docking System for docking and use the Boeing Lightweight Ablator for its heat shield. The Starliner's solar cells will provide more than 2.9 kW of electricity, and will be placed on top of the micro-

The New Era of Space Stations

meteoroid debris shield located at the bottom of the spacecraft's service module.

Unlike earlier U.S. space capsules, Starliner will make airbag-cushioned landings on the ground rather than into water. Five landing areas are planned in the Western United States, which will give the Starliner about 450 landing opportunities every year.

Starliner includes one space tourist seat, and the Boeing contract with NASA allows Boeing to price and sell passage to low Earth orbit on that seat.

The New Era of Space Stations

The New Era of Space Stations

4.3 USA-NASA Orion

Orion (officially Orion Multi-Purpose Crew Vehicle or Orion MPCV) is a class of partially reusable space capsules to be used in NASA's human spaceflight programs. The spacecraft consists of a Crew Module (CM) manufactured by Lockheed Martin and the European Service Module (ESM) manufactured by Airbus Defense and Space.

Capable of supporting a crew of six beyond low Earth orbit, Orion can last up to 21 days undocked and up to six months docked. It is equipped with solar panels, an automated docking system, and glass cockpit interfaces modeled after those used in the Boeing 787 Dreamliner. A single AJ10 engine provides the spacecraft's primary propulsion, while eight R-4D-11 engines, and six pods of custom reaction control system engines developed by Airbus, provide the spacecraft's secondary propulsion. Although compatible with other launch vehicles, Orion is primarily designed to launch atop a Space Launch System (SLS) rocket, with a tower launch escape system.

The New Era of Space Stations

Orion was originally conceived by Lockheed Martin as a proposal for the Crew Exploration Vehicle (CEV) to be used in NASA's Constellation program. Lockheed Martin's proposal defeated a competing proposal by Northrop Grumman, and was selected by NASA in 2006 to be the CEV. Originally designed with a service module featuring a new "Orion Main Engine" and a pair of circular solar panels, the spacecraft was to be launched atop the Ares I rocket. Following the cancellation of the Constellation program in 2010, Orion was heavily redesigned for use in NASA's Journey to Mars initiative; later named Moon to Mars. The SLS replaced the Ares I as Orion's primary launch vehicle, and the service module was replaced with a design based on the European Space Agency's Automated Transfer Vehicle. A development version of Orion's CM was launched in 2014 during Exploration Flight Test-1, while at least four test articles have been produced. As of 2020, three flight-worthy Orion spacecraft are under construction, with an additional one ordered, for use in NASA's Artemis program; the first of these is due to be launched in 2021 on Artemis 1. (This flight was delayed to 2022)

The Orion crew module (CM) is a reusable transportation capsule that provides a habitat for the crew, provides storage for consumables and research instruments, and contains the docking port for crew transfers. The crew module is the only part of the spacecraft that returns to Earth after each mission and is a 57.5° truncated cone shape with a blunt spherical aft end, 5.02 meters (16 ft 6 in) in diameter and 3.3 meters (10 ft 10 in) in length, with a mass of about 8.5 metric tons (19,000 lb). It was manufactured by the Lockheed Martin Corporation at Michoud Assembly Facility in New Orleans. It will have 50% more volume than the Apollo capsule and will carry

The New Era of Space Stations

four to six astronauts. After extensive study, NASA has selected the Avcoat ablator system for the Orion crew module. Avcoat, which is composed of silica fibers with a resin in a honeycomb made of fiberglass and phenolic resin, was formerly used on the Apollo missions and on the Space Shuttle orbiter for early flights.

Orion's CM will use advanced technologies, including:

Glass cockpit digital control systems derived from those of the Boeing 787.

An "autodock" feature, like those of Progress, the Automated Transfer Vehicle, and Dragon 2, with provision for the flight crew to take over in an emergency. Prior US spacecraft have all been docked by the crew.

Improved waste-management facilities, with a miniature camping-style toilet and the unisex "relief tube" used on the Space Shuttle.

A nitrogen/oxygen (N2/O2) mixed atmosphere at either sea level (101.3 kPa or 14.69 psi) or reduced (55.2 to 70.3 kPa or 8.01 to 10.20 psi) pressure.

Far more advanced computers than on prior crew vehicles.

The CM will be built of aluminum-lithium alloy. The reusable recovery parachutes will be based on the parachutes used on both the Apollo spacecraft and the Space Shuttle Solid Rocket Boosters, and will be constructed of Nomex cloth. Water landings will be the exclusive means of recovery for the Orion CM.

To allow Orion to mate with other vehicles, it will be equipped with the NASA Docking System. The spacecraft

The New Era of Space Stations

will employ a Launch Escape System (LES) along with a "Boost Protective Cover" (made of fiberglass), to protect the Orion CM from aerodynamic and impact stresses during the first 2 1/2 minutes of ascent. Its designers claim that the MPCV is designed to be 10 times safer during ascent and reentry than the Space Shuttle. The CM is designed to be refurbished and reused. In addition, all of Orion's component parts have been designed to be as modular as possible, so that between the craft's first test flight in 2014 and its projected Mars voyage in the 2030s, the spacecraft can be upgraded as new technologies become available.

As of 2019, the Spacecraft Atmospheric Monitor is planned to be used in the Orion CM

Orion Control Panels

The New Era of Space Stations

4.4 USA-Sierra Nevada-Dream Chaser

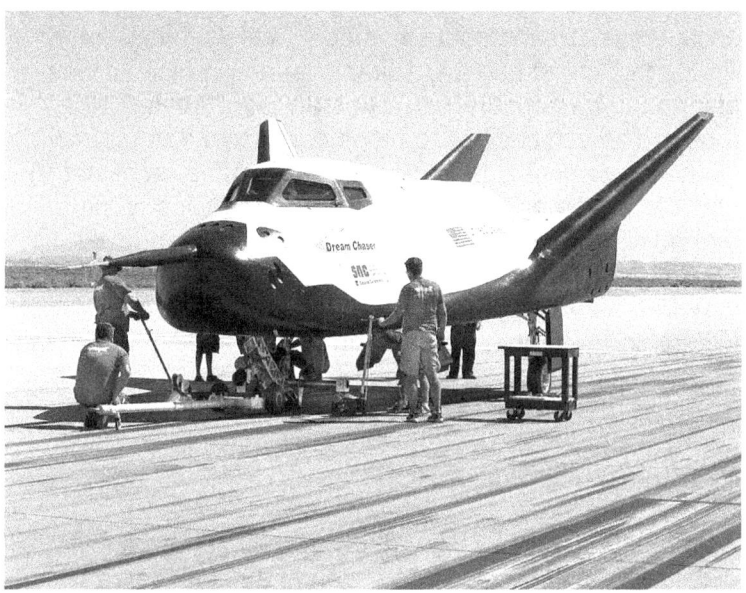

The Dream Chaser Cargo System is an American reusable lifting body spaceplane being developed by Sierra Nevada Corporation (SNC) Space Systems. Originally intended as a crewed vehicle, the Dream Chaser Space System, to be produced after the cargo variant is operational, is capable of carrying up to seven people and cargo to and from low Earth orbit.

The cargo Dream Chaser will resupply the International Space Station with both pressurized and unpressurized cargo. It will be launched vertically on the Vulcan Centaur rocket, and autonomously land horizontally on conventional runways. A proposed version operated by ESA would launch on Ariane 5.

The originally planned Dream Chaser Space System is a human-rated version designed to carry from two to seven

The New Era of Space Stations

people and cargo to orbital destinations such as the International Space Station. It was to have a built-in launch escape system and could fly autonomously if needed. Although it could use any suitable launch vehicle, it was planned to be launched on a human-rated Atlas V 412 rocket. The vehicle was to be able to return from space by gliding (typically experiencing less than 1.5 g on re-entry) and landing on any airport runway that handles commercial air traffic. Its reaction control system thrusters burned ethanol-based fuel, which is not an explosively volatile material, nor toxic like hydrazine, allowing the Dream Chaser to be handled immediately after landing, unlike the Space Shuttle. Its thermal protection system (TPS) was made up of silica-based tiles and a new composite material called Toughened Unipiece Fibrous Reusable Oxidation Resistant Ceramic (TUFROC).

As of 2020, the Sierra Nevada Corporation says it still plans to produce a crewed version of the spacecraft within the next 5 years. The company says it "never stopped working" on the crewed version and fully intends to launch it after the cargo version.

The New Era of Space Stations

5.0 Space Stations Being Planned Now

Here's how four space companies aim to replace the ISS. NASA is investing in Blue Origin, Nanoracks, and Northrop Grumman as they develop competing designs for a next-generation space station. The Pioneer Station is another recent design too.

An illustration of Blue Origin's Orbital Reef design, proposed for a new international space station.

The agency is investing some $416 million combined in the three companies to develop their designs, which include ISS-like modules or inflatable habitats. All would have to allow for future additional modules to be docked, Lego-style. NASA's financial contribution amounts to less than 40 percent of the total funding for the detailed designs, with the rest coming from private sources. Ultimately, the agency will choose only one of these plans to build.

"This is really the beginning of a new era. We did commercial crew, commercial cargo, and now commercial

The New Era of Space Stations

space stations. This is the next big step," says Marshall Smith, senior vice president of space systems at Nanoracks and a former deputy associate administrator at NASA.

NASA officials hope the ISS will continue operating at least until the late 2020s, when the first modules of the new station could launch. They're planning for a two-phase process. Until 2025, these companies will flesh out their blueprints in coordination with the space agency. Then in the second phase, NASA officials will choose one of the company's plans as the design they'll move forward with. Within two or three years, that company will launch its first module, which will provide accommodations for at least two astronauts to conduct research and experiments.

This will allow for a "seamless transition" from the ISS, Angela Hart, manager of NASA's commercial low-Earth orbit development program at Johnson Space Center in Houston, said at a press conference on Thursday. "This strategy will provide services the government needs at a lower cost to enable the agency to focus on its Artemis missions to the moon and on to Mars."

An artists' rendition of the space station design proposed by Northrop Grumman.

The New Era of Space Stations

Northrop Grumman is the most established company of the trio receiving awards; it dates back to the 1930s and has a longstanding relationship with NASA. Its proposal features a space station that appears most similar to the ISS, and it would use technologies and hardware that are mostly already available. It includes a cylindrical module, similar to the Habitation and Logistics Outpost, or HALO, which the company is already developing for NASA's planned Gateway space station that will orbit the moon. It will also include a larger version of its Cygnus cargo spacecraft, which has already been deployed multiple times to transport supplies to the ISS.

"We are trying to give NASA the option of something very reliable, something technically sound, and something we can do very quickly," says Rick Mastracchio, the company's director of business development for human exploration.

Nanoracks's proposed Starlab station will look completely different. Its large, inflatable habitat would have about a third the amount of pressurized cabin space as the ISS, and along with a science lab, docking port, power and propulsion element, and robotic arm, it could be propelled into orbit on a single launch. The Houston-based company is collaborating with Voyager Space (Nanoracks's majority shareholder) and Lockheed Martin.

While inflatable habitats are newer than metallic ones, the technology has existed for decades. Bigelow Aerospace's inflatable BEAM module has been docked at the ISS since 2016. The materials Nanoracks's habitat is made from are proprietary, but they're designed to offer protection from space radiation and debris, which continue to pose hazards, Smith says. "With inflatable technology, there's multiple layers it has to go through, layers that absorb the energy, like a Kevlar vest," Smith says.

An illustration of the Starlab station proposed by Nanoracks.

Blue Origin's Orbital Reef space station, which is in development with Sierra Space, includes both kinds of technologies: metallic core and science modules, as well as an inflatable habitat called LIFE. The architecture is designed to be a "mixed-use space business park" to support a variety of activities.

At any of these space stations, NASA will be the "anchor tenant," Mastracchio says. But as the commercial space travel market grows, the station will host other visitors, which could include those coming for tourism, sports, entertainment, and advertising. In fact, how the ISS's successor takes shape and which additional modules get

The New Era of Space Stations

prioritized for development could depend on market forces. In practice, that could initially create competition for the limited space available: Astronauts from the US, Europe, Russia, Japan, and Canada could end up vying for legroom and space for their research-focused experiments while private customers do the same for their activities.

But as the station is built up over time, different kinds of activities will be spread out through the various modules, so no one's sleeping in the lab, and tourists who just want to enjoy the view and zero-G life won't be in the way of the astronauts. "The easiest thing to imagine is essentially a dormitory, where all the habitation functions, like exercising and eating and socializing and sleeping, occur separately from laboratory functions or manufacturing functions," Brent Sherwood, senior vice president of advanced development programs at Blue Origin, said at the press conference.

But to have the first stages of a new space station in orbit by the late 2020s, NASA and its commercial partners have their work cut out for them. "NASA faces significant challenges with fully executing the plan in time to meet its 2028 goal and avoid a gap in availability of a low-Earth orbit destination," states the agency's Office of Inspector General report, published on November 30. The ISS costs about a third of NASA's annual human spaceflight budget. It's currently slated for retirement in 2024, but agency officials expect that date to be extended until 2030. In the meantime, astronauts will have to monitor cracks and leaks in hopes that the ISS remains safe until new modules start coming up.

These three new contracts fall under NASA's Commercial Low-Earth Orbit Development Program. Axiom Space's

The New Era of Space Stations

modules, designed for research and other applications, do as well. These include a habitation module, planned for launch in the second half of 2024, and lab and observatory modules. They're designed to connect to the ISS, and when the station finally retires they'll detach and become a free-flying commercial station.

Ultimately, NASA's competition could yield more than one winner, Jeffrey Manber, president of international and space stations for Voyager Space and chair of the board at Nanoracks, argued at the press conference: "At the end of this decade, there will be multiple privately owned space stations, maybe in different orbits."

The New Era of Space Stations

5.1 Orbital Reef Space Station

Here is more information on the Orbital Reef as described on the company website:

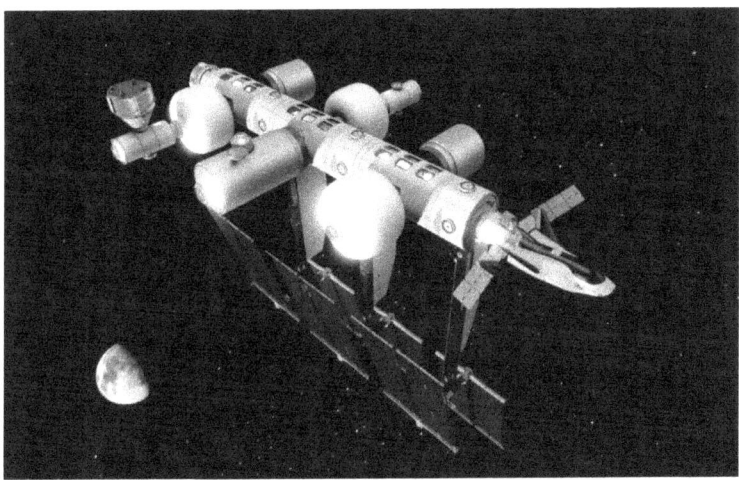

Orbital Reef will be the premier mixed-use space station in low Earth orbit for commerce, research, and tourism by the end of this decade

Low Earth Orbit, Unbound

Designed for use cases and ideas never before possible, Orbital Reef unlocks LEO by reducing cost and complexity for new kinds of customers. We provide end-to-end services, standard interfaces, and technical support needed by space flight novices: planning, payload development, training, transportation, data analysis, and security for your people or payloads (or both).

The New Era of Space Stations

A Revolution in Space Travel and Commerce

Commercially developed, owned, and operated, Orbital Reef opens doors to new markets and catalyzes the growth of a vibrant space ecosystem. It provides an "address on orbit" for use, lease, or ownership that is international and open to all.

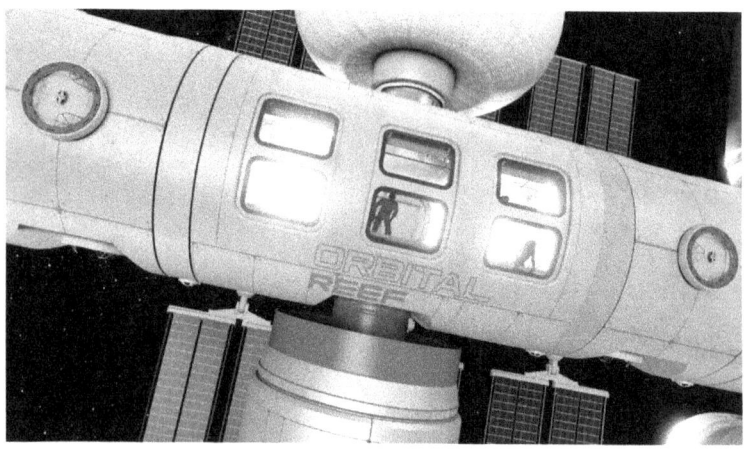

Reimagine Living and Working in Space

Orbital Reef introduces a new type of space architecture. Spacious modules with large Earth-facing windows let travelers take in the beauty of our planet and experience the thrill of weightlessness in complete comfort. Distinct quarters for living and working, and large hatches make a safe, sensible, and inspiring environment.

You will find habitation amenities for any length of visit within our orbital community, supported by medical care and recreation opportunities.

The New Era of Space Stations

Next-Generation Space Technology

Our research facilities are equipped to accelerate future discoveries and development of new industries. Seamless logistics, advanced robotics, and the Single Person Spacecraft make operations and excursions routine. The Dream Chaser spaceplane, Starliner spacecraft, and New Glenn launch system provide the backbone of cost-effective transportation.

WORKING ON ORBITAL REEF

Limitless horizons, the freedom of microgravity, the vacuum of space, and over-flying all of Earth's mid-latitudes offer unique commercial opportunities for a breathtaking range of applications and new markets like media, entertainment, and manufacturing.

A Mixed-Use Business Park

Whether film-making in microgravity, opening a space hotel, or conducting cutting edge research, Orbital Reef will lease locations that scale to fit your vision.

We sell only the utilities and services you need to sustain your business: power, cooling, high-bandwidth communications, information and physical security, robotic servicing, technician attention, stowage, and logistics.

The New Era of Space Stations

Out-of-This-World Research Facilities

Orbital Reef provides everything researchers need for fundamental science and applications development: experiment accommodations inside and outside; a spacious laboratory equipped with next-generation shared facilities; technician time; and proprietary provisions, even dedicated, closed-hatch modules. Whether for physical, biological, or Earth science, for new product development, or for testing exploration systems, you'll find it easier than ever to do your work in orbit.

The New Era of Space Stations

SPACE TOURISM

Anyone Can Experience Space
Experiencing space flight, and seeing the Earth from space, are life-changing. Floating free and weightless, awed by Earth's thin blue biosphere, you see sixteen vibrant sunrises and sunsets a day while flying over all of humanity.

We craft flight plans, training, and activities to take your adventure to a new level, whether for a short visit or a longer stay.

The New Era of Space Stations

The New Era of Space Stations

5.2 Northrop Grumman's Space Station

Northrop Grumman Corporation (NYSE: NOC), signed a Space Act Agreement with NASA under the Commercial Low-Earth Orbit Development program for $125.6 million to design a safe, reliable and cost-effective commercial free-flying space station in low Earth orbit (LEO). Northrop Grumman's commercial space station design will use current flight systems and advanced crew-focused technology under development that allows for rapid deployment with modular expansion to meet the growing needs of the space economy.

Northrop Grumman's low Earth orbit commercial free-flying space station design leverages flight proven elements to provide the base module for extended capabilities including science, tourism, industrial experimentation, and building of infrastructure beyond initial design.

"Under this agreement, the Northrop Grumman team will deliver a free flying space station design that is focused on commercial operations to meet the demands of an expanding LEO market," said Steve Krein, vice president, civil and commercial space, Northrop Grumman. "Our station will enable a smooth transition from International Space Station-based LEO missions to sustainable commercial-based missions where NASA does not bear all the costs, but serves as one of many customers."

This Space Act Agreement will enable Northrop Grumman to provide a detailed commercialization, operations and capabilities plan as well as space station requirements, mission success criteria, risk assessments, key technical and market analysis requirements, and preliminary design activities.

To support this effort, Northrop Grumman is building a team with unique capabilities and expertise, which includes Dynetics, a wholly owned subsidiary of Leidos, with other partners to be announced in the coming months.

Building on Northrop Grumman's commercial spacecraft experience with the Cygnus spacecraft and the Mission Extension Vehicle (MEV) as well as the in-production Habitation and Logistics Outpost (HALO), Northrop Grumman's design utilizes an overlapping stage approach that minimizes initial costs, provides revenue to offset subsequent development, and allows later capabilities to be added according to market needs.

The station will have the ability to support four permanent crewmembers initially, with plans to expand to an eight-person crew and further capability beyond that. The station is designed for a permanent presence of 15 years.

Northrop Grumman's design, using flight-proven elements, provides the base modules for commercial capabilities including science, tourism and manufacturing. Multiple docking ports will allow future expansion to support exploration crew analog habitats, laboratories, crew airlocks and facilities capable of artificial gravity.

Northrop Grumman is a technology company, focused on global security and human discovery. Our pioneering solutions equip our customers with capabilities they need to connect, advance and protect the U.S. and its allies. Driven by a shared purpose to solve our customers' toughest problems, our 90,000 employees define possible every day.

The New Era of Space Stations

The New Era of Space Stations

5.3 Nanoracks Space Station

Nanoracks will first be launched to connect to the International Space Station and then will disconnect to operate on its own.

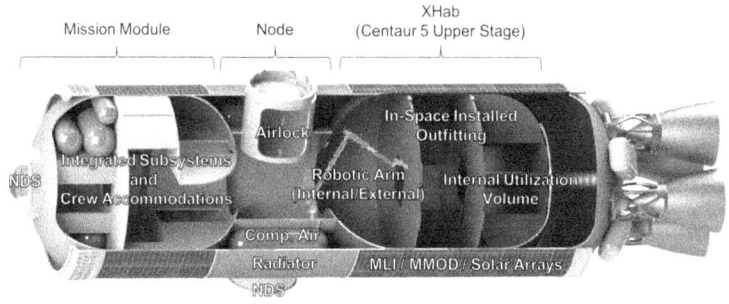

STARLAB-The First Ever Free-flying Commercial Space Station

Nanoracks spent the last decade mastering the commercial operation of space stations, meeting customer demand, charting market growth, and self-investing in private hardware on the ISS. Based on this experience, we are excited to introduce Starlab, a continuously crewed commercial platform supporting a business designed to enable science, research, and manufacturing for customers around the world.

The main structure of Starlab consists of a large inflatable habitat to be built by Lockheed Martin and a metallic docking node. Regenerative ECLSS enables continuous presence of four crew. They will live and work in a spacious 340m3 volume. The station also features a 60kW power and propulsion element, a large robotic arm for servicing cargo and external payloads, and a state-of-the-

The New Era of Space Stations

art laboratory system to host advanced research, science, and commercial capability.

The George Washington Carver (GWC) Science Park, the first science park in space, forms the core of Starlab. Named after the great American scientist, the GWC Science Park was founded to honor his legacy of scientific discovery for the benefit of life on Earth. Nanoracks and our experienced team of utilization specialists are tailoring the GWC Science Park to meet the needs of researchers and commercial customers. The Park features four main operational departments: a biology lab, plant habitation lab, physical science and materials research lab, and an open workbench area.

Nanoracks recently announced the unification of its commercial laboratory ecosystem on the ISS under the GWC Science Park name. The GWC serves as first ever in-space member organization of the International Association of Science Parks (IASP).

Nanoracks owns and operates Starlab and GWC Science Park. Nanoracks, a leader in commercial space services, will respond quickly to customer needs and prioritize on-orbit activities.

Nanoracks operates at the forefront of the commercial space station market with 10+ years of lessons learned as the largest commercial user of the ISS.

The New Era of Space Stations

One Launch

Starlab will deploy from one single launch in 2027, and offer an Initial Operating Capability (IOC) that allows for immediate and continuous crew presence and commercial activity.

Sustainable Business Model

Starlab operates a sustainable commercial business from a diverse customer base. Like the Bishop Airlock, Starlab is funded by private capital and welcomes NASA as one of many customers.

Significant Private Capital

Voyager Space, a leader in space exploration and New Space investment, will back Starlab development.

Benefit to Taxpayers

With NASA participating as one of many Starlab customers, the Agency receives comparable research capability and volume to the ISS, but at significantly lower construction and operational costs. This enables NASA to invest further in missions to the Moon, Mars, and beyond.

STEM Engagement

Starlab aligns with National Space Council priorities, providing STEM education activity through core partnerships with space education leader DreamUp and other key organizations dedicated to inclusivity and student engagement in the United States and globally.

The New Era of Space Stations

5.4 Pioneer Space Station

California Startup Aims to Build Space Hotel With Artificial Gravity by 2025

An illustration of the Pioneer-class space station, which is designed to have five modules built around a gravity ring.

Orbital Assembly Corporation announced plans to develop a space business park, complete with artificial gravity, that's designed to accommodate 28 guests in five modules built around a rotating gravity ring.

The California startup is aiming to make its first Pioneer-class space station operational by 2025, in what is an ambitious and likely unrealistic timeline. That said, Orbital Assembly is intent on making this the first commercial, hybrid space station that can be leveraged for both research and leisure.

The Pioneer station is one of two designs for commercial space stations currently being developed by the company,

The New Era of Space Stations

the first being the Voyager Station announced back in 2021. However, Pioneer is meant to precede Voyager, a larger undertaking that will be built with the aim of it being a sort of luxury space hotel capable of accommodating 400 guests at a time.

In an emailed statement, Rhonda Stevenson, chief executive officer of Orbital Assembly, said the Pioneer design is "a safe, secure, and reliable modular station that will generate revenue and profitability from both the tourist and commercial sectors sooner than our competitors who are adhering to NASA timetables."

The privately owned space stations are meant to take the place of the International Space Station (ISS), an orbiting laboratory that currently accommodates astronauts from NASA and the European Space Agency, as well as cosmonauts from Russia's Roscosmos. However, the current plan is to retire the ISS in 2030 by plunging it into a remote region of the Pacific Ocean known as Point Nemo. The public-private hybrid space stations commissioned by NASA are meant to take the place of ISS in low Earth orbit, where they will serve both research and commercial interests.

Private space company Axiom Space is also planning to build its own commercial space station, one designed to house future visitors and host scientific research. Axiom is hoping to launch the first section of the station to low Earth orbit in 2024.

Despite the pending competition, Orbital Assembly is hoping to become the dark horse of commercial space tourism. "Multiple revenue streams from commercial, research and tourism markets will enable us to subsidize the travel market for a one to two-week stay," Stevenson

The New Era of Space Stations

said. "While launch costs continue to be a barrier, we expect tourists will be motivated to plan shorter, or more frequent, stays as space travel becomes less expensive." The orbiting Pioneer outposts are designed to simulate one-sixth of Earth's gravity, which they'll do by spinning around a gravity ring that measures about 200 feet (61 meters) in diameter. Having a weightless environment on a space station would allow people to still move around "while eating or drinking out of a cup normally and sleeping without having to be attached to a bed," according to the company's statement. Creating artificial gravity is also a way to mitigate the detrimental health effects of microgravity on the human body.

Orbital Assembly previously announced that it would start building its larger Voyager space station by 2025 and that it would become operational in 2027. That timeline has been met with skepticism from the industry, but the recent statement from the company suggests Pioneer will be up and running before Voyager is in orbit.

In early February, the company, which runs on private investment, announced $1 million in newly raised funding, which allowed for a slew of new hires. It will likely need to acquire more money if it hopes to build an orbiting space station; for reference, NASA allocated a total of $415.6 million for the three contracts awarded to Blue Origin, Nanoracks, and Northrop Grumman.

Space tourism is a tempting and potentially lucrative business that's attracting a lot of corporate attention right now. This nascent industry is still taking shape, however, and the challenges of actually building out a flying space hotel remain unknown, as is the willingness of people to cough up millions for a ticket to space.

The New Era of Space Stations

The New Era of Space Stations

6.0 New Technologies for Space Stations

Many of the technologies developed for life support on the International Space Station will be used in the new space stations.

These included water recycling, data management and communications, solar panels for electricity, and thermal management heat pipes.

In this chapter we will discuss a few new technologies which have either been approved for new usage or have been developed but not used on a large scale yet.

6.1 Inflatable Structures

Bigelow Aerospace has done the most work on inflatable structures for space. Here is some of what they created:

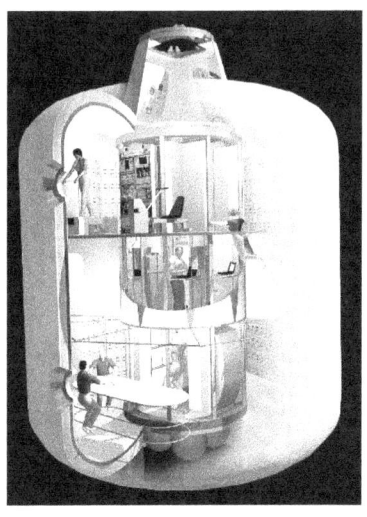

The New Era of Space Stations

Bigelow Aerospace is an American aeronautics and outer space technology company which manufactures and develops expandable space station modules. Bigelow Aerospace was founded by Robert Bigelow in 1998, and is based in North Las Vegas, Nevada. It is funded in large part by the profit Bigelow gained through his ownership of the hotel chain, Budget Suites of America.

By 2013, Bigelow had invested US$250 million in the company. Bigelow stated on a number of occasions that he was prepared to fund Bigelow Aerospace with about US$500 million through 2015 in order to achieve launch of full-scale hardware.

Bigelow Aerospace announced in 2010 that they intended to create a modular set of space habitats for creating or expanding space stations but despite many concepts and models, never completed a working space station beyond two small prototypes that flew in 2006 and 2007.

In March 2020, the company laid off all 88 of its employees due to the COVID-19 pandemic, and planned to rehire staff when conditions permitted.

Bigelow originally licensed the multi-layer, expandable space module technology from NASA in 2000 after Congress canceled the International Space Station (ISS) TransHab project following delays and budget constraints in the late 1990s.

Bigelow has three Space Act agreements whereby Bigelow Aerospace is the sole commercializer of several of NASA's key expandable module technologies.

Bigelow continued to develop the technology for a decade, redesigning the module fabric layers – including adding

The New Era of Space Stations

proprietary extensions of Vectran shield fabric, "a double-strength variant of Kevlar" – and developing a family of uncrewed and crewed expandable spacecraft in a variety of sizes. Bigelow invested US$75 million in proprietary extensions to the NASA technology by mid-2006, and US$180 million into the technology by 2010.

By 2010, Robert Bigelow had invested US$180 million in the company, which by 2013 had grown to US$250 million of his personal fortune. Bigelow stated on multiple occasions that he was prepared to fund Bigelow Aerospace with up to about US$500 million through 2015 in order to achieve launch of full-scale hardware.

In early 2010, NASA came full circle to once again investigate "making inflatable space-station modules to make roomier, lighter, cheaper-to-launch spacecraft" by announcing plans in its budget proposal released 22 February 2010. NASA considered connecting a Bigelow expandable craft to the ISS for safety, life support, radiation shielding, thermal control and communications verification testing for the next three years", and in December 2012, signed a US$17.8 million contract with Bigelow to develop the Bigelow Expandable Activity Module (BEAM), then projected to fly in 2015. The module was berthed to the International Space Station on 16 April 2016, and was inflated on 28 May 2016. As of January 2022 it remains at the station.

Since early on, Bigelow has been intent on "pursuing markets for a variety of users including biotech and pharmaceutical companies and university research, entertainment applications and government military and civil users". The business model includes "'leasing out' small space stations or habitats made of one or more [B330] inflatable modules for different research

The New Era of Space Stations

communities or corporations". Despite these broad plans for space commercialization, the space tourism destination and space hotel monikers were frequently used by many media outlets following the 2006/2007 launches of Genesis I and Genesis II. Robert Bigelow has been explicit that he is aiming to do business in space in a new way, with "low cost and rapid turnaround, contrary to traditional NASA ISS and Space Shuttle operations and bureaucracy".

In October 2010, Bigelow announced that it had agreements with six sovereign nations to utilize on-orbit facilities of the commercial space station: UK Astronomy Technology Centre (United Kingdom), Netherlands Space Office (Netherlands), Defense South Australia (Australia), Singapore Government Technology Development Agency (Singapore), Japan Manned Space Systems Corporation, chairman is a previous director of JAXA (Japan) and Swedish National Space Board (Sweden). In February 2011, Dubai of the United Arab Emirates became the seventh nation to have signed on.

In 2011, Bigelow employed an in-house team of model makers, coming from the film and architecture industries, to make detailed models of their space habitats and space stations. Scale models were sent to "potential customers, including governments and corporations, as a reminder of the possibilities".

The New Era of Space Stations

6.2 Growing Food in Space

The ability to grow food in space both reduces what needs to be sent up from Earth as expensive payloads, but it also provides fresh food and will be important to have as man reaches our further from the Earth.

Plants in space

The growth of plants in outer space has elicited much scientific interest. In the late 20th and early 21st century, plants were often taken into space in low Earth orbit to be grown in a weightless but pressurized controlled environment, sometimes called space gardens. In the context of human spaceflight, they can be consumed as food and/or provide a refreshing atmosphere. Plants can metabolize carbon dioxide in the air to produce valuable oxygen, and can help control cabin humidity. Growing plants in space may provide a psychological benefit to human spaceflight crews. Usually the plants were part of studies or technical development to further develop space gardens or conduct science experiments. To date plants taken into space have had mostly scientific interest, with only limited contributions to the functionality of the spacecraft, however the Apollo Moon tree project was more or less forestry inspired mission and the trees part of a country's bicentennial celebration.

The first challenge in growing plants in space is how to get plants to grow without gravity. This runs into difficulties regarding the effects of gravity on root development, providing appropriate types of lighting, and other challenges. In particular, the nutrient supply to root as well as the nutrient biogeochemical cycles, and the microbiological interactions in soil-based substrates are particularly complex, but have been shown to make possible space farming in hypo- and micro-gravity.

NASA plans to grow plants in space to help feed astronauts, and to provide psychological benefits for long-term space flight. In 2017, aboard ISS in one plant growth device, the 5th crop of Chinese cabbage (Brassica rapa) from it included an allotment for crew consumption, while the rest was saved for study. An early discussion of plants in space, were the trees on the brick moon space station, in the 1869 short story "The Brick Moon".

History

In the 2010s there was an increased desire for long-term space missions, which led to desire for space-based plant production as food for astronauts. An example of this is vegetable production on the International Space Station in Earth orbit. By the year 2010, 20 plant growth experiments had been conducted aboard the International Space Station.

Several experiments have been focused on how plant growth and distribution compares in micro-gravity, space conditions versus Earth conditions. This enables scientists to explore whether certain plant growth patterns are innate or environmentally driven. For instance, Allan H. Brown tested seedling movements aboard the Space Shuttle Columbia in 1983. Sunflower seedling movements were recorded while in orbit. They observed that the seedlings still experienced rotational growth and circumnation despite lack of gravity, showing these behaviors are instinctual.

Other experiments have found that plants have the ability to exhibit gravitropism, even in low-gravity conditions. For instance, the ESA's European Modular Cultivation System enables experimentation with plant growth; acting as a miniature greenhouse, scientists aboard the International Space Station can investigate how plants react in variable-gravity conditions. The Gravi-1 experiment (2008) utilized the EMCS to study lentil seedling growth and amyloplast movement on the calcium-dependent pathways. The results of this experiment found that the plants were able to sense the direction of gravity even at very low levels. A later experiment with the EMCS placed 768 lentil seedlings in a centrifuge to stimulate various gravitational changes; this experiment, Gravi-2 (2014), displayed that plants change calcium signaling towards root growth while being grown in several gravity levels.

Many experiments have a more generalized approach in observing overall plant growth patterns as opposed to one specific growth behavior. One such experiment from the Canadian Space Agency, for example, found that white spruce seedlings grew differently in the anti-gravity space environment compared with Earth-bound seedlings; the space seedlings exhibited enhanced growth from the shoots and needles, and also had randomized amyloplast distribution compared with the Earth-bound control group.

Early efforts

The first organisms in space were "specially developed strains of seeds" launched to 134 km (83 mi) on 9 July 1946 on a U.S. launched V-2 rocket. These samples were not recovered. The first seeds launched into space and successfully recovered were maize seeds launched on 30 July 1946. Soon followed rye and cotton. These early suborbital biological experiments were handled by Harvard University and the Naval Research Laboratory and were concerned with radiation exposure on living tissue. On September 22 1966, Kosmos 110 launched with two dogs and moisturized seeds. Several of those seeds germinated, the first to do so, resulting in lettuce, cabbage and some beans that had greater yield than their controls on Earth. In 1971, 500 tree seeds (Loblolly pine, Sycamore, Sweetgum, Redwood, and Douglas fir) were

The New Era of Space Stations

flown around the Moon on Apollo 14. These Moon trees were planted and grown with controls back on Earth where no changes were detected.

Space station era

In 1982, the crew of the Soviet Salyut 7 space station conducted an experiment, prepared by Lithuanian scientists (Alfonsas Merkys and others), and grew some Arabidopsis using Fiton-3 experimental micro-greenhouse apparatus, thus becoming the first plants to flower and produce seeds in space. A Skylab experiment studied the effects of gravity and light on rice plants. The SVET-2 Space Greenhouse successfully achieved seed to seed plant growth in 1997 aboard space station Mir. Bion 5 carried Daucus carota and Bion 7 carried maize (aka corn).

Plant research continued on the International Space Station. Biomass Production System was used on the ISS Expedition 4. The Vegetable Production System (Veggie) system was later used aboard ISS. Plants tested in Veggie before going into space included lettuce, Swiss chard, radishes, Chinese cabbage and peas. Red Romaine lettuce was grown in space on Expedition 40 which were harvested when mature, frozen and tested back on Earth.

Expedition 44 members became the first American astronauts to eat plants grown in space on 10 August 2015, when their crop of Red Romaine was harvested. Since 2003 Russian cosmonauts have been eating half of their crop while the other half goes towards further research. In 2012, a sunflower bloomed aboard the ISS under the care of NASA astronaut Donald Pettit. In January 2016, US astronauts announced that a zinnia had blossomed aboard the ISS.

The New Era of Space Stations

In 2017 the Advanced Plant Habitat was designed for ISS, which was a nearly self-sustaining plant growth system for that space station in low Earth orbit. The system is installed in parallel with another plant grown system aboard the station, VEGGIE, and a major difference with that system is that APH is designed to need less upkeep by humans. APH is supported by the Plant Habitat Avionics Real-Time Manager. Some plants that were to be tested in APH include Dwarf Wheat and Arabidopsis. In December 2017 hundreds of seeds were delivered to ISS for growth in the VEGGIE system.

In 2018 the Veggie-3 experiment at the ISS, was tested with plant pillows and root mats. One of the goals is to grow food for crew consumption. Crops tested at this time include cabbage, lettuce, and mizuna. In 2018, the PONDS system for nutrient deliver in microgravity was tested.

The Seedling Growth series of experiments to study the mechanisms of tropisms and the cell/cycle were performed on the ISS between 2013 and 2017. These experiments also involved using the model plant Arabidopsis thaliana, and were a collaboration between NASA (John Z. Kiss as PI) and ESA (F. Javier Medina as PI).

On 30 November 2020, astronauts aboard the ISS collected the first harvest of radishes grown on the station. A total of 20 plants was collected and prepared for transportation back to Earth. There are currently plans to repeat the experiment and grow a second batch.

The New Era of Space Stations

6.3 Working in Space

There will also be new technologies to work in space which are safer than today's space suits and which allow the pilot to stay out in space longer to work.

Genesis Engineering is one of the companies devising these tiny spaceships and they are a partner on the Orbital Reef Team.

Here is a description of their innovative single person spacecraft:

Benefits of a Single-Person Spacecraft for Weightless Operations (Stop Walking and Start Flying)

Historically, less than 20 percent of crew time related to extravehicular activity (EVA) is spent on productive external work. For planetary operations space suits are still the logical choice; however, for safe and rapid access to the weightless environment, spacecraft offer compelling

The New Era of Space Stations

advantages. FlexCraft, a concept for a single-person spacecraft, enables anytime access to space for short or long excursions by different astronauts. For the International Space Station (ISS), going outside is time-consuming, requiring pre-breathing, donning a fitted space suit, and pumping down an airlock. For each ISS EVA this is between 12.5 and 16 hours. FlexCraft provides immediate access to space because it operates with the same cabin atmosphere as its host.

Furthermore, compared to the space suit pure oxygen environment, a mixed gas atmosphere lowers the fire risk and allows use of conventional materials and systems. For getting to the worksite, integral propulsion replaces hand-overhand translation or having another crew member operate the robotic arm. This means less physical exertion and more time at the work site. Possibly more important, in case of an emergency, FlexCraft can return from the most distant point on ISS in less than a minute. The one-size-fits-all FlexCraft means no on-orbit inventory of parts or crew time required to fit all astronauts. The shirtsleeve cockpit uses conventional displays and controls which means the work is not strenuous, there is no suit trauma and rest days are not required. Furthermore, there is no need to collect hand tools because manipulators are equipped with force multiplying end-effectors that can deliver the precise torque for the job.

FlexCraft is an efficient solution for asteroid exploration allowing all crew to use one vehicle with no risk of contamination. And, because FlexCraft is a vehicle, its design offers better radiation and micro-meteoroid protection than space suits.

The New Era of Space Stations

7.0 New Space Station Usage Plans

There are many ideas on what commercial activities can be done on the new commercial stations. Some are already making money and others we don't know for sure.

7.1 Space Tourism

Space Tourism is starting today and will only grow as orbital "hotels" become available for tourists.

A handful of companies – including one publicly traded name – are competing neck and neck to be leaders in the emerging market.

But what space tourism entails, and how much it costs per person, varies greatly depending on a company's technological capabilities. For example, both Virgin Galactic and SpaceX expect to fly private paying passengers to space next year. But, while passengers flying with both companies would go to space by the Federal Aviation Administration's definition, a Virgin Galactic passenger spends about 0.04% as much time in space as on a SpaceX trip, while a ride with Elon Musk's company is expected to cost roughly 200 times as much.

Whether a passenger reaches suborbital and orbital space is the major difference in the destinations of the human spaceflight offerings in development. Because of that difference, there are notable distinctions in the cost, experience and even risk of what it means to be a space tourist.

UBS in a report last year estimated that space tourism, with both suborbital and orbital together, has a potential market value of $3 billion by 2030. More recently, space industry consultancy Northern Sky Research broke out its

The New Era of Space Stations

expectations for suborbital versus orbital tourism. By 2028, NSR expects suborbital will be a $2.8 billion market, with $10.4 billion in total revenue over the next decade, while orbital will be a $610 million market, with $3.6 billion in total revenue over the next decade.

Here's how the small but growing market breaks down, and which companies are involved in each area.

There are two companies competing in the realm of suborbital tourism: Virgin Galactic, which debuted on the public market last year and trades under the ticker "SPCE," and Blue Origin, the private space company funded almost entirely by Amazon founder Jeff Bezos.

Both of the companies' systems are rocket-powered and capable of carrying up to six passengers on a flight, but that is where the similarities end.

Virgin Galactic's spacecraft SpaceShipTwo, which has two pilots in addition to the passengers, is docked underneath a jet-powered carrier aircraft known as WhiteKnightTwo. With the spacecraft attached, the carrier aircraft takes off from a runway and climbs to an altitude of more than 40,000 feet. Then the spacecraft is dropped, free-falling briefly before firing its rocket motor and ascending to an altitude of about 295,000 feet, or roughly 90 kilometers. The spacecraft essentially does a slow back flip at the edge of space, with passengers spending a few minutes floating in microgravity, before it re-enters and then glides back to land on its runway in New Mexico. The company reuses the spacecraft, replacing the hybrid rocket engine and reconnecting it to the carrier aircraft.

Blue Origin's more traditional rocket New Shepard launches with a domed capsule on top of the about 60 foot tall booster. It ascends straight up, with the capsule separating near the top of the flight and reaching an altitude of more than 330,000 feet, or about 100 kilometers. There the capsule floats for a few minutes in microgravity before returning back to Earth, slowing down using a system of parachutes to land in the West Texas desert floor.

The New Era of Space Stations

But unlike conventional rockets, New Shepard's booster also comes back to land separately – with the company reusing the boosters for future launches.

Virgin Galactic has sold tickets to about 600 passengers at a price between $200,000 and $250,000 each, although the company expects it could increase its prices substantially for the first commercial flights. Blue Origin has said its ticket pricing is yet to be determined, but Bezos expects his company will price flights on New Shepard comparable to competitors.

To date Virgin Galactic has flown five people to space on two test flights. All five are company employees, with four pilots controlling the spacecraft and chief astronaut trainer Beth Moses riding along as a test passenger on the second flight. Richard Branson flew on this ship in July 2021.

Jeff Bezos also flew to space on his rocket in July 2021.

Additionally, Virgin Galactic has said that passengers will spend three days training before a flight, while Blue Origin expects its passengers will train for just one day.

Given the price point of the flights, both Virgin Galactic and Blue Origin are targeting high net worth individuals for the suborbital experiences. Virgin Galactic chief space officer George Whitesides has previously categorized space tourism flights as an "out-of-home luxury experience," which is the fastest-growing part of the luxury market. And, when Blue Origin also begins flying passengers, Whitesides said he thinks both companies will have more than enough demand for flights.

The New Era of Space Stations

"Globally, we think around 2 million people can experience this over the coming years at this price point. Over time, we'll be able to reduce that price point and at that point the market just explodes. It's 10 times as many at 40 million people," Whitesides said last year.

Cowen and UBS have each recently conducted surveys of high net worth individuals and their interest in suborbital tourism. Cowen's survey exceeded Whitesides' estimation, as the firm found suborbital flights have a total addressable market of about 2.4 million people among individuals with a net worth of more than $5 million. UBS surveyed more than 6,000 high net worth individuals specifically on flying with Virgin Galactic. About 20% of those UBS surveyed said they are "likely to purchase a ticket on a spacecraft within 1 year" of the company beginning regular flights. That number increases to between more than 35% "after several years of safe operation," UBS said.

Additionally, UBS highlighted how much Virgin Galactic's space tourism market expands as the price comes down. The firm estimated that there are about 1.78 million people with a net worth of more than $10 million but there are about twice that many with a net worth between $5 million and $10 million – and about 37.1 million people with a net worth between $1 million and $5 million.

Orbital tourism

Unlike suborbital, which reaches an altitude of about 100 kilometers (or 330,000 feet) and gives passengers a few minutes in space, orbital missions reach an altitude of over 400 kilometers (or 1.3 million feet) and spend days or even more than a week in space. To date, orbital space tourism has largely been limited to a few flights to the International Space Station that used Russian Soyuz spacecraft.

The New Era of Space Stations

But SpaceX, with its Falcon 9 rocket and Crew Dragon capsule, has now entered the orbital tourism arena. This summer SpaceX launched and returned two NASA astronauts with its spacecraft for the first time ever, in a mission that was historic for both Musk's company and the U.S. space agency. The test flight made SpaceX the first private company to send people to orbit, a feat only previously achieved by government superpowers.

In 2021 and 2022 there have also been several strictly tourism flights into orbit and to the ISS.

The SpaceX launch system is similar to Blue Origin's, but with a more powerful rocket and a larger capsule. Its Crew Dragon spacecraft is built to hold as many as seven passengers and sits on top of the company's 230 foot tall Falcon 9 rocket booster. Launching from NASA's Kennedy Space Center in Florida, it takes the spacecraft several hours to reach either the ISS or its intended orbit. NASA astronauts on the recent Demo-2 mission described riding in SpaceX's capsule as "a little bit smoother" than the Space Shuttles of the past, which were "a little bit rougher, at least at the beginning."

The New Era of Space Stations

SpaceX

Even before that Demo-2 mission launched, SpaceX had already signed two separate agreements with companies looking to fly private paying passengers to space. While NASA's astronauts spend months up at the ISS, the private missions will be only about 10 days at most. While SpaceX hasn't disclosed specifically how much each of the current contracts are worth, previously announced contracts mean that it will likely cost about $50 million per person to fly with Crew Dragon.

In addition to the launch costs, a 10-day mission to the ISS would rack up a $350,000 bill with NASA. Under the agency's cost structure unveiled last year, NASA would get $35,000 a night per person, as compensation for the agency's services a tourist would need while on board the ISS.

The New Era of Space Stations

But Crew Dragon likely won't be the only option for private passengers to get to the ISS in the coming years. While testing delays mean the spacecraft remains in development, Boeing's Starliner capsule is also designed to carry as many as seven passengers. And, under Boeing's contract with NASA to fly four astronauts at a time, the company is allowed to sell the fifth seat to prospective space tourists. A Boeing spokesperson told CNBC on Friday that the company has a team actively looking to sell that fifth seat.

Finally, Crew Dragon and Starliner are likely to remain the two best options for orbital tourists, SpaceX is also working on its next-generation Starship rocket. It's the company's top priority, as Musk wants to build a fully reusable rocket system that can launch cargo or as many as 100 people at a time. But, while SpaceX has signed a deal to fly Japanese billionaire Yusaku Maezawa around the moon with Starship in 2023, Musk noted earlier this month that the rocket will have to complete "hundreds of missions with satellites before we put people on board."

Starship prototypes have launched and landed in short flight tests this past year but the rocket has yet to reach orbit.

Orbital brokers and services

In addition to the handful of companies building rockets and spacecraft that fly people, there are a few that also help find interested passengers and get them ready to launch. Space Adventures, Axiom Space and Virgin Galactic each offer some variation of orbital space tourism services, although the companies don't build or launch spacecraft that go to orbit.

The New Era of Space Stations

Over the past two decades U.S.-based Space Adventures has flown seven tourists using Russian spacecraft. At a reported cost of more than $20 million per person, the private clients typically spent over a week on board the ISS. Most recently, Space Adventures signed one of the two deals announced with SpaceX. Using a Crew Dragon capsule, Space Adventures plans to have SpaceX fly four tourists on a "free-flyer" mission to orbit, meaning they won't dock with the ISS but instead will orbit the Earth for five days before returning.

Houston-based start-up Axiom Space signed the other deal with SpaceX, to fly a professionally-trained commander and three passengers to the ISS in October 2021 using Crew Dragon. It was a 12 day mission, with two days of travel and ten days on board the space station. In this case Axiom acts as more than just a broker, providing all the services necessary – from training to management and more.

NASA in May 2021 announced that it is working with actor Tom Cruise to film a movie on the ISS. And, while the agency did not confirm Cruise is flying with Axiom or SpaceX, Musk tweeted that the mission "should be a lot of fun!" On Wednesday, space industry publication Spaceflight Now reported that Cruise will be one of the passengers on an Axiom mission.

Axiom also confirmed this month that it is working with a U.S. television production company called Space Hero for another 10 day trip to the ISS. Scheduled for 2023, Space Hero said it plans send the winning contestant of a reality TV show to the space station, with Axiom training the crew and managing the mission.

The New Era of Space Stations

Finally, while Virgin Galactic doesn't plan to develop an orbital spacecraft, the company signed an agreement with NASA that sets the company "up to become a player in the provision of that service," it said. Since signing the agreement, Virgin Galactic said it now has "deposit agreements" for orbital spaceflights with 12 customers, although it has yet to finalize pricing.

"We expect pricing to be competitive with other offerings in the market," Whitesides told investors during the company's most recent quarterly earnings call.

The New Era of Space Stations

7.2 Additional Space Station Activities

Many of the new commercial space stations plan to lease space to either government agencies or other companies who want to work in space.

These leased spaces could be for manufacturing, science experimentation, or even military imaging of the Earth.

Things which make sense to manufacture in space are those which benefits from zero gravity or vacuum.

This includes different types of biological processes, manufacturing alloys of materials which don't mix well on Earth, growing crops in space, and other types of work.

Images from space stations of what is on Earth also make sense. NASA has been using spacecraft and the ISS for decades to shoot images of Earth for many reasons.

Experiments on the ISS for more than twenty years have shown that zero gravity or even the vacuum of space provides conditions which are very difficult or impossible to reproduce on Earth.

The New Era of Space Stations

Another focus would be as a base to conduct other Space construction activities like building Solar Power plants, building space ships to visit other planets, and things we haven't even thought of yet.

The New Era of Space Stations

8.0 The Deep Space Gateway

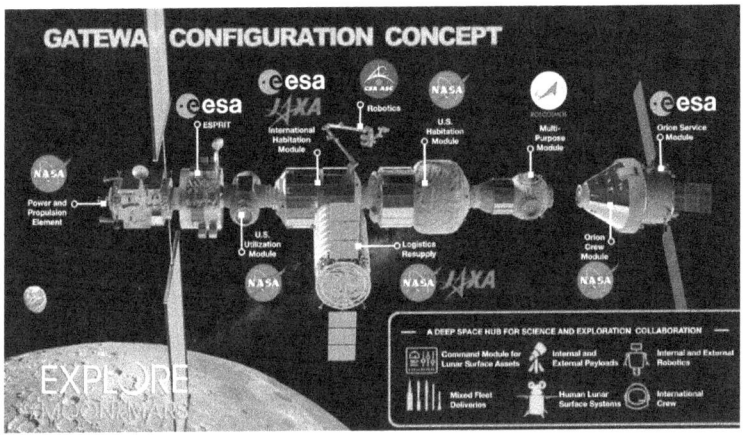

The Gateway will be an international effort and as of fall 2019 over twenty five countries want to sign up to participate.

The Lunar Gateway is designed to provide a way station for trips to land on the Moon. It will be in an elongated orbit around the Moon positioned such that landing craft can leave the gateway to land on different parts of the Moon from different orbital positions of the Gateway.

Lunar Orbit Details:

The Lunar Gateway is planned to be deployed in a highly elliptical seven-day near-rectilinear halo orbit (NRHO) around the Moon, which would bring the station within 3,000 km (1,900 mi) of the lunar north pole at closest approach and as far away as 70,000 km (43,000 mi) over the lunar south pole. Traveling to and from cislunar space (lunar orbit) is intended to develop the knowledge and experience necessary to venture beyond the Moon and into deep space.

The New Era of Space Stations

The proposed NRHO orbit would allow lunar expeditions from the Gateway to reach a low polar orbit with a delta-v of 730 m/s and a half a day of transit time. Orbital station-keeping would require less than 10 m/s of delta-v per year, and the orbital inclination could be shifted with a relatively small delta-v expenditure, allowing access to most of the lunar surface. Spacecraft launched from Earth would perform a powered flyby of the Moon (delta-v = ~180 m/s) followed by a ~240 m/s delta-V NRHO orbit insertion burn to dock with the Gateway as it approaches the apoapsis point of its orbit. The total travel time would be 5 days; the return to Earth would be similar in terms of trip duration and delta-V requirement if the spacecraft spends 11 days at the Gateway. The crewed mission duration of 21 days and ~840 m/s delta-V are limited by the capabilities of the Orion life support and propulsion systems.

Lunar Gateway Modules are planned as follows:

Contracted modules

A) The **Power and Propulsion Element (PPE)** started development at the Jet Propulsion Laboratory during the now canceled Asteroid Redirect Mission. The original concept was a robotic, high performance solar electric spacecraft that would retrieve a multi-ton boulder from an asteroid and bring it to lunar orbit for study. When ARM was cancelled, the solar electric propulsion was repurposed for the Gateway. The PPE will allow access to the entire lunar surface and act as a space tug for visiting craft. It will also serve as the command and communications center of the Gateway. The PPE is intended to have a mass of 8-9 tons and the capability to generate 50 kW of solar electric power for its ion thrusters, which can be supplemented by chemical propulsion. It is currently planned to launch on a commercial launch vehicle in 2022. In May 2019,

The New Era of Space Stations

Maxar Technologies was contracted by NASA to manufacture this module, which will also supply the station with electrical power and is based on Maxar's 1300 series satellite bus. The PPE will use Advanced Electric Propulsion System (AEPS) Hall-effect thrusters. Maxar was awarded a firm-fixed price contract of $375 million to build the PPE. NASA is supplying the PPE with an S-band communications system to provide a radio link with nearby vehicles and a passive docking adapter to receive the Gateway's future utilization module.

B) The **Habitation and Logistics Outpost (HALO),** also called the Minimal Habitation Module (MHM) and formerly known as the Utilization Module, will be built by Northrop Grumman Innovation Systems (NGIS). A commercial launch vehicle would launch the HALO before the end of year 2023. The HALO is based on a Cygnus Cargo resupply module to the outside of which radial docking ports, body mounted radiators (BMRs), batteries and communications antennae will be added. The HALO will be a scaled-down habitation module, yet, it will feature a functional pressurized volume providing sufficient command, control & data handling capabilities, energy storage and power distribution, thermal control, communications and tracking capabilities, two axial and up to two radial docking ports, stowage volume, environmental control and life support systems to augment the Orion spacecraft and support a crew of four for at least 30 days.

C) The **European System Providing Refueling, Infrastructure and Telecommunications (ESPRIT) service module** will provide additional xenon and hydrazine capacity, additional communications equipment, and an airlock for science packages. It will have a mass of approximately 4 tons (8,800 lb), and a length of 3.91 m (12.8 ft). The studies and design are being performed mostly by Airbus and OHB. The module construction was approved in November 2019.

D) The **International Habitation Module (iHAB)** will be an additional habitation module built by ESA in collaboration with Japan. Together with the HALO module, they will provide a combined 125 m3 (4,400 cu ft) of habitable volume to the station.

Proposed modules

The concept for the Lunar Gateway is still evolving, and these modules have also been proposed to be added to the design:

The **Gateway Logistics Modules** will be used to refuel, resupply and provide logistics on board the space station. The first logistics module sent to the Gateway will also arrive with a robotic arm, which will be built by the Canadian Space Agency.

The New Era of Space Stations

The **Gateway Airlock Module** will be used for performing extravehicular activities outside the space station and would have the docking port for the proposed Deep Space Transport.

The New Era of Space Stations

The New Era of Space Stations

9.0 Latest Moon Base Plans

I would not normally include moon bases in this book but they will also use the same life support technologies pioneered on the International Space Station. This is also going to be an early habitat off the Earth.

NASA's Artemis Base Camp on the Moon Will Need Light, Water, Elevation

American astronauts in 2024 will take their first steps near the Moon's South Pole: the land of extreme light, extreme darkness, and frozen water that could fuel NASA's Artemis lunar base and the agency's leap into deep space.

Scientists and engineers are helping NASA determine the precise location of the Artemis Base Camp concept. Among the many things NASA must take into account in choosing a specific location are two key features: The site must bask in near continuous sunlight to power the base and moderate extreme temperature swings, and it must offer easy access to areas of complete darkness that hold water ice.

The New Era of Space Stations

While the South Pole region has many well-illuminated areas, some parts see more or less light than others. Scientists have found that at some higher elevations, such as on crater rims, astronauts would see longer periods of light. But the bottoms of some deep craters are shrouded in near constant darkness, since sunlight at the South Pole strikes at such a low angle it only brushes their rims.

These unique lighting conditions have to do with the Moon's tilt and with the topography of the South Pole region. Unlike Earth's 23.5-degree tilt, the Moon is tilted only 1.5 degrees on its axis. As a result, neither of the Moon's hemispheres tips noticeably toward or away from the Sun throughout the year as it does on Earth — a phenomenon that gives us sunnier and darker seasons here. This also means that the height of the Sun in the sky at the lunar poles doesn't change much during the day. If a person were standing on a hilltop near the lunar South Pole during daylight hours, at any time of year, they would see the Sun moving across the horizon, skimming the surface like a flashlight laying on a table.

What future growth of the Moonbase might look like

The New Era of Space Stations

"It's such a dramatic terrain down there," said W. Brent Garry, a geologist at NASA's Goddard Space Flight Center in Greenbelt, Maryland. Garry is working with engineers on a virtual reality tour of the Moon's South Pole to help immerse astronauts, scientists, and mission planners in the exotic environment of that region as they prepare for a human return to the Moon.

While a base camp site will require lots of light, it is also important for astronauts to be able to take short trips into permanently dark craters. Scientists expect that these shadowed craters are home to reservoirs of frozen water that explorers could use for life support. "One idea is to set up camp in an illuminated zone and traverse into these craters, which are exceptionally cold," said NASA Goddard planetary scientist Daniel P. Moriarty, who's involved with NASA's South Pole site analysis and planning team. Temperatures in some of the coldest craters can dip to about -391 degrees Fahrenheit (-235 degrees Celsius).

Initial plans include landing a spacecraft on a relatively flat part of a well-lit crater rim or a ridge. "You want to land in the flattest area possible, since you don't want the landing vehicle to tip over," Moriarty said.

The landing area, ideally, should be separated from other base camp features — such as the habitat or solar panels — by at least half a mile, or 1 kilometer. It also ought to be situated at a different elevation to prevent descending spacecraft from spraying high-speed debris at equipment or areas of scientific interest. Some scientists have estimated that as a spacecraft thrusts its engines for a soft landing, it could potentially spray nearly a million pounds, or hundreds of thousands of kilograms, of surface particles, water, and other gases across the surface.

The New Era of Space Stations

"You want to take advantage of the landforms, such as hills, that can act as barriers to minimize the impact of contamination," says Ruthan Lewis, a biomechanical and industrial engineer, architect, and a leader on NASA's South Pole site analysis and planning team. "So, we're looking at distances, elevations, and slopes in our planning."

Preparing to explore the surface of the Moon goes well beyond designing and building safe spacecraft and spacesuits. NASA also has to ensure the surface vehicles and suits have the mobility required to do science, and that astronauts have the tools they need to identify and scoop up rock and soil samples.

At the Moon, it's critical to keep the area around the landing site and base camp as pristine as possible for scientists. For instance, among the many interesting features of the South Pole region is its location right between the Earth-facing side of the Moon, or the near side, and the side we never see from Earth, known as the far side.

These two hemispheres are geologically very different, with the far side more heavily cratered and its crust thicker than on the near side. Scientists don't know why the two sides formed this way.

The Artemis Base Camp has to be on the Earth-facing side to make it easier for engineers to use radio waves to communicate with astronauts working on the Moon. But scientists expect that over billions of years of meteorite impacts to the Moon's surface, rocks, and dust from each hemisphere were kicked up and strewn about the other, so it's possible that astronauts could collect samples of the far side from their base camp on the near side.

The New Era of Space Stations

10.0 Russian Space Station Plans

Due to the United States and Russia being on opposing sides of the war in Ukraine, the Russians have announced they will be leaving the International Space Station in the near future. Here are some of the Russian plans for their future space stations:

The Russian Orbital Service Station is a proposed Russian orbital space station scheduled to begin construction in 2025. Initially an evolution of the Orbital Piloted Assembly and Experiment Complex (OPSEK) concept, ROSS developed into plans for a new standalone Russian space station built from scratch without modules from the Russian Orbital Segment of the ISS.

In April 2021, Roscosmos (the Russian government space agency) officials announced plans to possibly exit from the International Space Station program after 2024, stating concerns about the condition of its aging modules.

The New Era of Space Stations

However, nothing is yet official. A new space station, named Russian Orbital Space Station, operated entirely by Roscosmos, would be launched starting in the mid-2020s. The first crewed mission is planned for 2026.

ROSS will operate at a 400 km altitude sun-synchronous orbit, which will allow it to monitor the entire surface of the Earth, especially the Arctic region. This orbit will enable the station to serve two important functions: high-frequency observations of Russia from space, and easier access to the station compared to the ISS, which will allow for more medical and physiological experiments to be conducted than what is currently feasible on the Russian Orbital Segment of the ISS

Planned modules

NEM-1, also known as Science Power Module 1 (SPM-1), will be the core module of ROSS. Initially intended to be launched to the International Space Station in 2024, NEM-1 will instead undergo 1.5–2 years of redesign to prepare the module for its new role as part of ROSS. As of April 2021, NEM-1 is scheduled to launch in 2025 on an Angara A5 launch vehicle from Vostochny Cosmodrome and new Core module (similar to NEM-1) is scheduled to launch in 2028.

ROSS is envisioned to include up to seven modules, with 2035 being the targeted completion date. The first stage of construction will consist of four modules: the base NEM-1 module, an upgraded NEM, a node module, and a gateway module. The second stage will include logistics and production modules, as well as a platform module for servicing spacecraft. A commercial module for up to four space tourists is also under consideration.

The New Era of Space Stations

11.0 Chinese Space Stations

China has a very active space program and has their own new space station too. Here we review their current space station and much bigger station plans for the future.

11.1 China's Current Space Station

Tiangong is a space station that the Chinese Manned Space Agency (CMSA) is building in low Earth orbit. In May 2021, China launched Tianhe, the first of the orbiting space station's three modules, and the country aims to finish building the station by the end of 2022. CMSA hopes to keep Tiangong inhabited continuously by three astronauts for at least a decade. The space station will host many experiments from both China and other countries.

Tiangong, which means "Heavenly Palace," will consist of Tianhe, the main habitat for astronauts, and two modules dedicated to hosting experiments, Mengtian and Wentian,

The New Era of Space Stations

both of which are due to launch in 2022. Shenzhou spacecraft, launching from Jiuquan in the Gobi Desert, will send crews of three astronauts to the space station, while Tianzhou cargo spacecraft will launch from Wenchang on the Chinese island of Hainan to deliver supplies and fuel to the station.

Tiangong will be much smaller than the International Space Station (ISS), with only three modules compared with 16 modules on the ISS. Tiangong will also be lighter than the ISS, which weighs about 400 tons (450 metric tons) following the recent addition of Russia's Nauka module.

The 54-foot-long (16.6 meter) Tianhe module launched with a docking hub that allows it to receive Shenzhou and Tianzhou spacecraft, as well as welcome the two later experiment modules. A large robotic arm will help position the Mengtian and Wentian modules and assist astronauts during spacewalks.

Tianhe is much larger than the Tiangong 1 and 2 test space labs China launched in the last decade and nearly three times heavier, at 24 tons (22 metric tons). The new Tiangong, visiting spacecraft and cargo spacecraft will expand the usable space for the astronauts; so much that they'll feel as though "they will be living in a villa," compared with how little space was available on previous Chinese space labs, Bai Linhou, deputy chief designer of the space station, told CCTV.

Tianhe features regenerative life support, including a way to recycle urine, to allow astronauts to stay in orbit for long periods. It is the main habitat for the astronauts and also houses the propulsion systems to keep the space station in orbit.

The New Era of Space Stations

China has said it will take 11 launches to finish Tiangong: three module launches, four crewed missions and four Tianzhou spacecraft to supply cargo and fuel. The first three launches — Tianhe, Tianzhou 2 and Shenzhou 12 — have gone smoothly.

Once completed, Tiangong will be joined by a huge, Hubble-like space telescope, which will share the space station's orbit and be able to dock for repairs, maintenance and possibly upgrades. Named Xuntian, which translates to "survey the heavens," the telescope will have a 6.6-foot (2 m) diameter mirror like Hubble but will have a field of view 300 times greater. Xuntian will aim to survey 40% of the sky over 10 years using its huge, 2.5-billion-pixel camera.

The space station could potentially be expanded to six modules, if everything goes according to plan. "We can further expand our current three-module space station combination into a four-module, cross-shaped combination in the future," Bai told CCTV. The second Tianhe core module could then allow two more modules to join the orbital outpost.

The New Era of Space Stations

The New Era of Space Stations

11.2 China's Plans for a Giant Space Station

China Wants to Build a Mega Spaceship That's Nearly a Mile Long. A proposal plans to study how to build a giant spacecraft

China is investigating how to build ultra-large spacecraft that are up to 0.6 mile (1 kilometer) long. But how feasible is the idea, and what would be the use of such a massive spacecraft?

The project is part of a wider call for research proposals from the National Natural Science Foundation of China, a funding agency managed by the country's Ministry of Science and Technology. A research outline posted on the foundation's website described such enormous spaceships as "major strategic aerospace equipment for the future use of space resources, exploration of the mysteries of the universe, and long-term living in orbit."

The New Era of Space Stations

The foundation wants scientists to conduct research into new, lightweight design methods that could limit the amount of construction material that has to be lofted into orbit, and new techniques for safely assembling such massive structures in space. If funded, the feasibility study would run for five years and have a budget of 15 million yuan ($2.3 million).

The project might sound like science fiction, but former NASA chief technologist Mason Peck said the idea isn't entirely off the wall, and the challenge is more a question of engineering than fundamental science.

"I think it's entirely feasible," Peck, now a professor of aerospace engineering at Cornell University, told Live Science. "I would describe the problems here not as insurmountable impediments, but rather problems of scale."

By far the biggest challenge would be the price tag, noted Peck, due to the huge cost of launching objects and materials into space. The International Space Station (ISS), which is only 361 feet (110 meters) wide at its widest point according to NASA, cost roughly $100 billion to build, Peck said, so constructing something 10 times larger would strain even the most generous national space budget.

Much depends on what kind of structure the Chinese plan to build, though. The ISS is packed with equipment and is designed to accommodate humans, which significantly increases its mass. "If we're talking about something that is simply long and not also heavy then it's a different story," Peck said.

The New Era of Space Stations

Building techniques could also reduce the cost of getting a behemoth spaceship into space. The conventional approach would be to build components on Earth and then assemble them like Legos in orbit, said Peck, but 3D-printing technology could potentially turn compact raw materials into structural components of much larger dimensions in space.

An even more attractive option would be to source raw materials from the moon, which has low gravity compared with Earth, meaning that launching materials from its surface into space would be much easier, according to Peck. Still, that first requires launch infrastructure on the moon and is therefore not an option in the short term.

BIG SPACESHIP, BIG PROBLEMS

A structure of such massive proportions will also face unique problems. Whenever a spacecraft is subjected to forces, whether from maneuvering in orbit or docking with another vehicle, the motion imparts energy to the spaceship's structure that causes it to vibrate and bend, Peck explained. With such a large structure, these vibrations will take a long time to subside so it's likely the spacecraft will require shock absorbers or active control to counteract those vibrations, he said.

Designers will also have to make careful trade-offs when deciding what altitude the spacecraft should orbit at, Peck said. At lower altitudes, drag from the outer atmosphere slows vehicles down, requiring them to constantly boost themselves back into a stable orbit. This is already an issue for the ISS, Peck noted, but for a much larger structure, which has more drag acting on it and would require more fuel to boost back into place, it would be a major concern.

The New Era of Space Stations

On the flip side, launching to higher altitudes is much more expensive, and radiation levels increase quickly the further from Earth's atmosphere an object gets, which will be a problem if the spacecraft houses humans.

But while building such a structure might be technically possible, it's not feasible in any practical sense, said Michael Lembeck, a professor of aerospace engineering at the University of Illinois at Urbana-Champaign who has worked on both government and commercial space programs.

"It's kind of like us talking about building the Starship Enterprise," he told Live Science. "It's fantastical, not feasible, and fun to think about, but not very realistic for our level of technology," given the cost, he said.

Given the research project's tiny budget, it is likely only meant to be a small, academic study to map out the very earliest contours of such a project and identify technological gaps, Lembeck said. For comparison, the budget to build a capsule to take astronauts to the ISS was $3 billion. "So the level of effort here is extremely small compared to the outcomes that are desired," he added.

There are also questions about what such a big spacecraft would be used for. Lembeck said possibilities include space manufacturing facilities that take advantage of microgravity and abundant solar power to build high-value products like semiconductors and optical equipment, or long-term habitats for off-world living. But both would entail enormous maintenance costs.

"The space station is a $3 billion a year enterprise," Lembeck added. "Multiply that for larger facilities and it quickly becomes a rather large, expensive enterprise to pull off."

China has also expressed interest in building enormous solar power arrays in orbit and beaming the power back to Earth via microwave beams, but Peck said the economics of such a project just don't stack up. Peck has done some back-of-the-envelope calculations and estimates it would cost around $1,000 per watt, compared with just $2 per watt for energy generated from solar panels on Earth.

Perhaps the most promising application for such a large space structure would be scientific, Peck said. A space telescope of that scale could potentially see features on the surface of planets in other solar systems. "That could be transformative for our understanding of extrasolar planets and potentially life in the universe," he added.

The New Era of Space Stations

The New Era of Space Stations

12.0 Future Advanced Space Colonies

In this chapter is a lot of information on the potential of future space colonies from my book "Designing and Building Space Colonies" to give you an idea of the potential growth of what we can build in space.

Over the last fifty years or so, many space habitat designs have been proposed and lots of engineers, scientists, and visionaries have worked on these concepts. None of them can be built with today's technologies, but they all offer incredible visions for the future.

The Rotating Wheel

Both scientists and science fiction writers have thought about the concept of a rotating wheel space station since the beginning of the 20th century. Konstantin Tsiolkovsky wrote about using rotation to create an artificial gravity in space in 1903. Herman Potočnik introduced a spinning wheel station with a 30-meter diameter in his Problem der Befahrung des Weltraums (The Problem of Space Travel). He even suggested it be placed in a geostationary orbit.

In the 1950s, Wernher von Braun and Willy Ley, writing in Colliers Magazine, updated the idea, in part as a way to stage spacecraft headed for Mars. They envisioned a rotating wheel with a diameter of 76 meters (250 feet). The 3-deck wheel would revolve at 3 RPM to provide artificial one-third gravity. It was envisaged as having a crew of 80. In 1959, a NASA committee opined that such a space station was the next logical step after the Mercury program.

The New Era of Space Stations

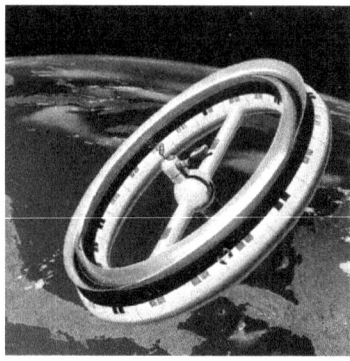

An internal cross section of what the inside of the wheel might look like:

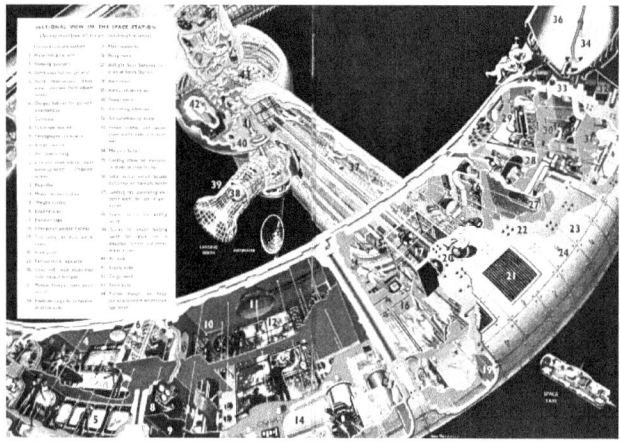

O'Neil Cylinders

The O'Neill cylinder (also called an O'Neill colony) is a space settlement design proposed by American physicist Gerard K. O'Neill in his 1976 book The High Frontier: Human Colonies in Space. O'Neill proposed the

The New Era of Space Stations

colonization of space for the 21st century, using materials extracted from the Moon and later from asteroids.

An O'Neill cylinder would consist of two counter-rotating cylinders. The cylinders would rotate in opposite directions in order to cancel out any gyroscopic effects that would otherwise make it difficult to keep them aimed toward the Sun. Each would be 5 miles (8.0 km) in diameter and 20 miles (32 km) long, connected at each end by a rod via a bearing system. They would rotate so as to provide artificial gravity via centrifugal force on their inner surfaces. Here is a look at a notional view of the inside of an O'Neil habitat:

The New Era of Space Stations

Two counter rotating O'Neil Habitats to stop habitat rotation:

A Close-up view of the outside of O'Neil habitats:

The New Era of Space Stations

The Stanford Torus

The Stanford torus is a proposed NASA design for a space habitat capable of housing 10,000 to 140,000 permanent residents.

The Stanford torus was proposed during the 1975 NASA Summer Study, conducted at American University, and Stanford University, with the purpose of exploring and speculating on designs for future space colonies (Gerard O'Neill later proposed his Island One or Bernal sphere as an alternative to the torus). "Stanford torus" refers only to this particular version of the design, as the concept of a ring-shaped rotating space station was previously proposed by Wernher von Braun and Herman Potočnik.

It consists of a torus, or doughnut-shaped ring, that is 1.8 km in diameter (for the proposed 10,000 person habitat described in the 1975 Summer Study) and rotates once per minute to provide between 0.9g and 1.0g of artificial gravity on the inside of the outer ring via centrifugal force. Sunlight is provided to the interior of the torus by a system of mirrors. The ring is connected to a hub via a number of "spokes", which serve as conduits for people and materials travelling to and from the hub. Since the hub is at the rotational axis of the station, it experiences the least artificial gravity and is the easiest location for spacecraft to dock. Zero-gravity industry is performed in a non-rotating module attached to the hub's axis.

The interior space of the torus itself is used as living space, and is large enough that a "natural" environment can be simulated; the torus appears similar to a long, narrow, straight glacial valley whose ends curve upward and eventually meet overhead to form a complete circle. The population density is similar to a dense suburb, with part of the ring dedicated to agriculture and part to housing.

The New Era of Space Stations

The Stanford Torus outside view:

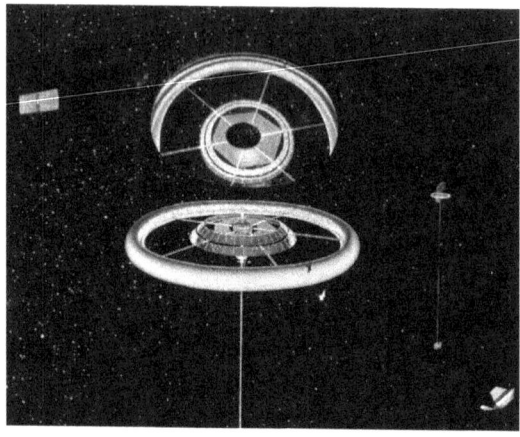

Stanford Torus inside view:

The New Era of Space Stations

The Bernal Sphere

In a series of studies held at Stanford University in 1975 and 1976 with the purpose of speculating on designs for future space colonies, Dr. Gerard K. O'Neill proposed Island One, a modified Bernal sphere with a diameter of only 500 m (1,600 ft.) rotating at 1.9 RPM to produce a full Earth artificial gravity at the sphere's equator.

The result would be an interior landscape that would resemble a large valley running all the way around the equator of the sphere. Island One would be capable of providing living and recreation space for a population of approximately ten thousand people, with a "Crystal Palace" habitat used for agriculture. Sunlight was to be provided to the interior of the sphere using external mirrors to direct it in through large windows near the poles. The form of a sphere was chosen for its optimum ability to contain air pressure and its optimum mass-efficiency at providing radiation shielding

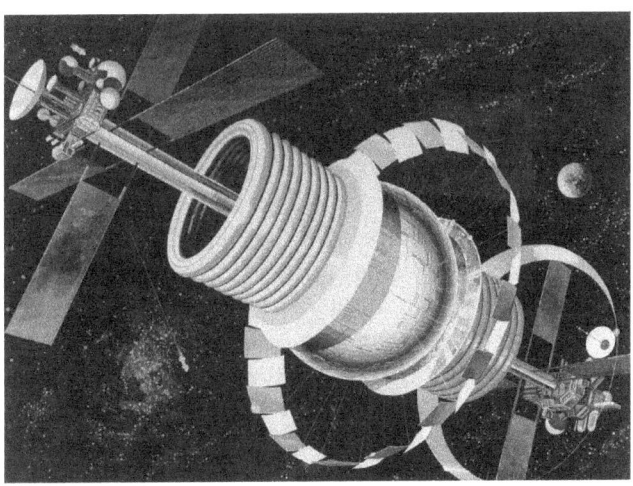

The New Era of Space Stations

The interior of a Bernal Sphere:

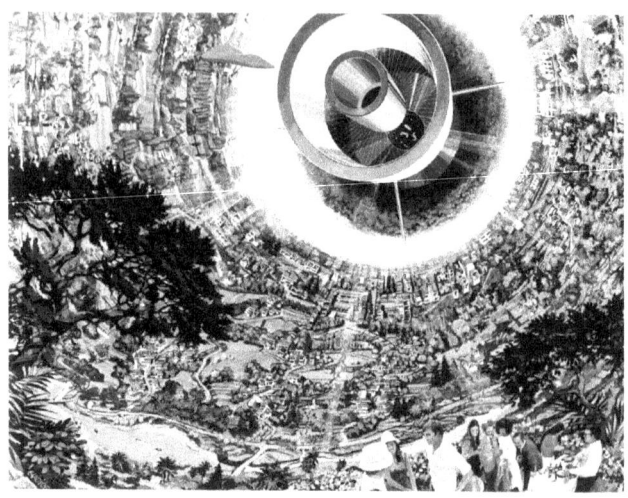

11.5 Kalpana One

Kalpana One is intended to be the first, and smallest, of a family of space settlements. The size is determined by the limited rotation rate humans are assumed to tolerate, 2*rpm*. The rotation rate drives the radius to achieve 1*g* pseudo-gravity, and the radius drives the length due to angular moment of inertia requirements. For later, larger settlements in the Kalpana family, the rotation rate may be reduced, increasing the radius and the allowable length. Kalpana One solves some of the problems found in earlier designs: excessive shielding mass, large appendages, lack of natural sunlight, rotational instability, lack of wobble control, and some catastrophic failure modes. Much is left to be done before a practical space settlement can be fully designed and built. Just as our distant ancestors left the warm oceans and colonized dry land, it is our task to settle the vast, empty reaches of space; thereby ensuring the survival and growth of civilization, humanity, and life itself. Let's get to work.

The New Era of Space Stations

Here are exterior and interior views:

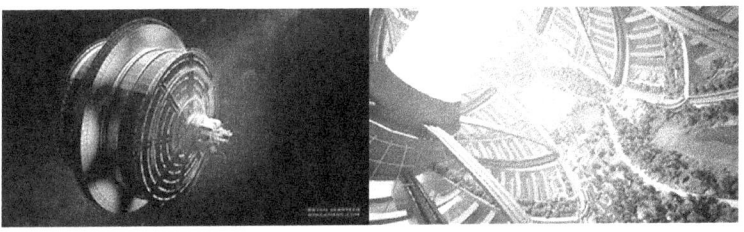

The New Era of Space Stations

The New Era of Space Stations

13.0 Summary

This book is just an update on the rapidly moving growth in Space enterprises. There is a lot happening now with new low cost space transportation being developed and lots of new projects in development for living and working in space.

I've written quite a few books on space and the topic moves so quickly that it is hard to keep up with it all.

But space travel and space technologies are a passion of mine and with all of the new opportunities to go into space I haven't given up hope of still getting there in my lifetime.

All the Best,

Martin K. Ettington
May 2022

The New Era of Space Stations

The New Era of Space Stations

14.0 Bibliography

Ettington, M. K. (2017). *Designing and Building Space Colonies.*

Ettington, M. K. (2019). *All About Moon Bases.*

Ettington, M. K. (2020). *Spaceships: Past, Present, and Future.*

https://genesisesi.com/wp-content/uploads/2019/05/Publication-Benefits-of-a-Single-Person-Spacecraft_Griffin_.pdf. (2022). Retrieved from Benefits of a Single Person Spacecraft.

https://gizmodo.com/orbital-assembly-space-hotel-artificial-gravity-1848855049. (2022). Retrieved from Pioneer Station.

https://nanoracks.com/starlab/. (2022). Retrieved from Starlab.

https://news.northropgrumman.com/news/releases/northrop-grumman-signs-agreement-with-nasa-to-design-space-station-for-low-earth-orbit. (2022). Retrieved from Northrop Grumman Signs Agreement with Nasa to Design a Space Station for Low Earth Orbit.

https://www.cnbc.com/2020/09/26/space-tourism-how-spacex-virgin-galactic-blue-origin-axiom-compete.html. (2020). Retrieved from Space Tourism Competitors.

https://www.orbitalreef.com/. (2022). Retrieved from Orbital Reef.

The New Era of Space Stations

https://www.scientificamerican.com/article/china-wants-to-build-a-mega-spaceship-thats-nearly-a-mile-long/. (2022). Retrieved from China wants to Build a Mega Spaceship thats nearly a mile long.

https://www.space.com/tiangong-space-station. (2022). Retrieved from The Tiangong Space Station.

https://www.wired.com/story/heres-how-3-space-companies-aim-to-replace-the-iss/. (2022). Retrieved from How 3 Space Comapnies Aim to Replace the ISS.

The New Era of Space Stations